Beginner's Guide to Astronomical Telescope Making

By the same author

THE SUN'S FAMILY
Weidenfeld & Nicolson, 1962

ASTRONOMY WITH BINOCULARS
Faber & Faber, 1963 & 1975

THE PAN BOOK OF ASTRONOMY
Pan Books, 1964 & 1972

SPACE INTRUDER
Hamish Hamilton, 1965

THE MOON-WINNERS
Hamish Hamilton, 1965

STARS AND PLANETS
Crowell (U.S.A.), 1965

THE EARTH'S NEIGHBOURS
Weidenfeld & Nicolson, 1968

THE AMATEUR ASTRONOMER'S HANDBOOK
Crowell (U.S.A.), 1968 & 1974

ASTRONOMY FOR AMATEURS
Cassell, 1969

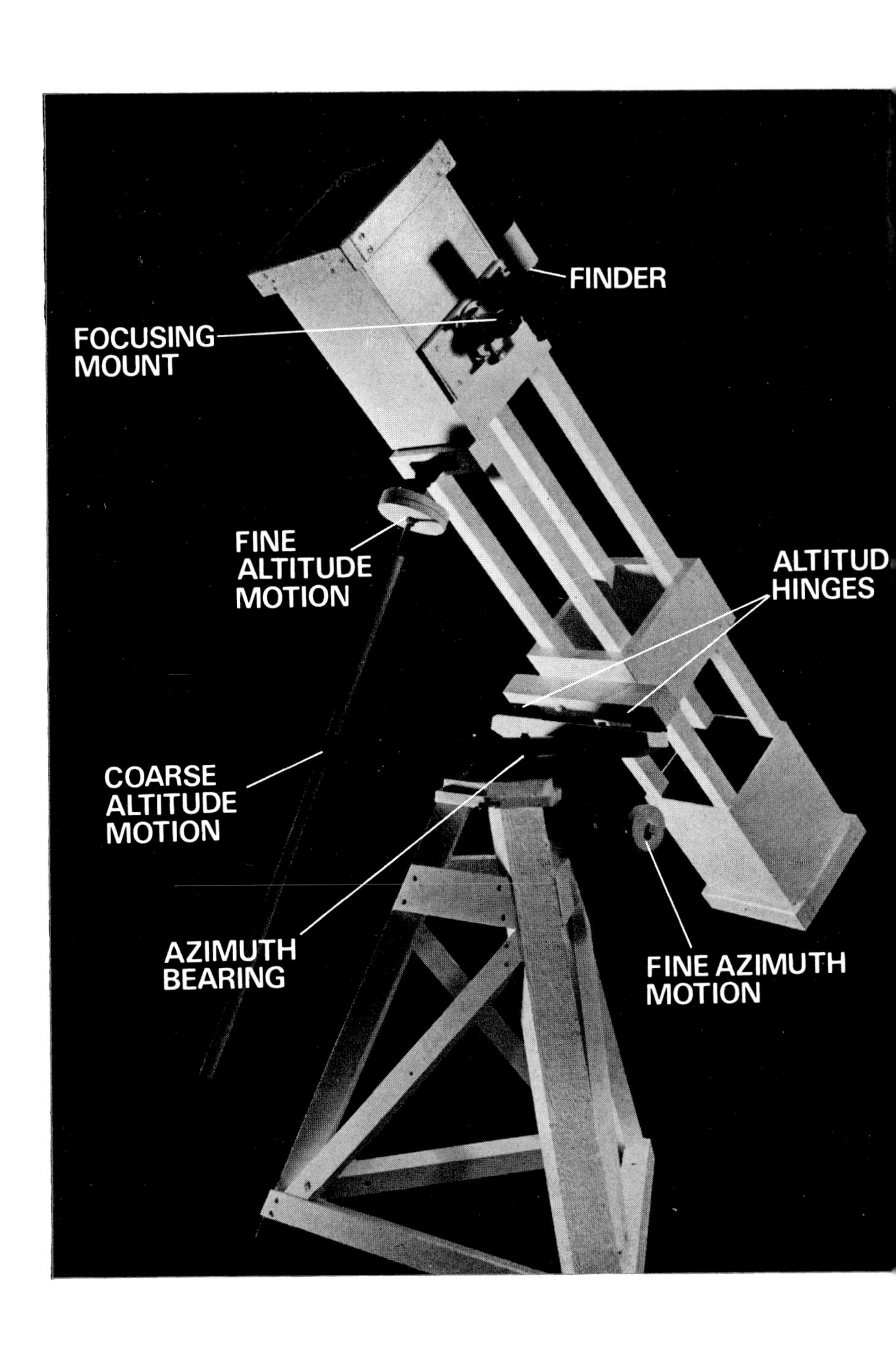
FINDER
FOCUSING MOUNT
FINE ALTITUDE MOTION
ALTITUD
HINGES
COARSE ALTITUDE MOTION
AZIMUTH BEARING
FINE AZIMUTH MOTION

James Muirden

Beginner's Guide to Astronomical Telescope Making

PELHAM BOOKS

TRANSATLANTIC ARTS, INC.
Sole Distributor For North America
P. O. Box 6086
ALBUQUERQUE, NM 87197 U.S.A.

First published in Great Britain by
PELHAM BOOKS LTD
52 Bedford Square
London WC1B 3EF
1975

ISBN 0 7207 0822 2

Set and printed in Great Britain by
Tonbridge Printers Ltd, Peach Hall Works, Tonbridge, Kent
in Times eleven on thirteen point on paper supplied by
P. F. Bingham Ltd, and bound by James Burn
at Esher, Surrey

Contents

Illustrations

PHOTOGRAPHS

FIGURES

Acknowledgements

My thanks are due to Steven Anderson and Peter B. York, both of whom read the manuscript of this book and suggested improvements, and to William Dillon for his labours over the illustrations. I am also most grateful to Dr James G. Baker for permission to use his object-glass design in Appendix B.

In more general terms I must record my indebtedness to two leading British opticians, H. E. Dall and E. J. Hysom, for affording me much help and encouragement over the years.

Preface

At the present time, there about ten thousand people in this country who are sufficiently interested in astronomy to have joined a national or local society. A poll among them would reveal an overwhelming majority having nothing larger than a pair of binoculars with which to conduct their hobby. When we remember that, for every society member, there must be half a dozen 'casual' amateurs who either do not yet feel confident enough to join a group, or else have no local society, we find that there must be many thousands of people wanting a powerful telescope to bring the stars and planets near enough for proper observations to be made.

An astronomical telescope is a highly specialised instrument, and the ultimate clue to its quality lies in the excellence of the mirrors or lenses it contains. Fine optical components cannot be mass-produced, and so good-quality instruments are expensive. Few people can part with well over a hundred pounds just to satisfy their hobby!

The idea of *making* a telescope has probably been no more than a passing fancy, to be instantly dismissed as totally impractical if the instrument is to be of reasonable size and power. The purpose of this book is to show that anyone who is even remotely good with his hands, and has patience enough to be a good astronomer – which is what it is all about – can make himself an excellent telescope for only a few pounds. He will also find himself entering the unexpectedly fascinating world of optical work, where molecules are molehills and microns are mountains, and wavelengths of light are no longer infinitesimal abstractions but represent very real amounts of glass that must be observed, acted upon, and brought to order. There is something majestic in the balance between the fundamentals of optical accuracy on the one hand, and the supreme vastness of the universe on the other; and the thought that one has laboured with eye and hand to such fine limits to reveal what is, on the opposite side of the scale, so unimaginably vast, gives an inspiration to the observer's work that must be lacking in the use of a manufactured instrument, however imposing and excellent it may be.

Indeed, optical work and telescope-making themselves constitute a hobby that may prove as absorbing as astronomical observation, and the fact that one fits naturally into the other makes it a source of endless surprise that so little telescope-making activity exists in this country. I do not believe for a moment that there is proportionately less native talent here than elsewhere, yet the fact remains that in the U.S.A. there are numerous telescope-making groups, while talented individuals are producing instruments of excellent quality. In this country there seems to be no organised or publicised work along these lines at all. The root cause of this sad situation is the lack of any regular exchange of information in the British astronomical journals. Neither will a visit to the library or bookshop be of much help. This book, as far as I know, is only the third volume to be published since the war by a British writer on the subject! Foreign books are available – though few in number – to help increase our sense of isolation.

My aim, then, has been to help the frustrated observer to make, in the shortest possible time, a powerful and effective astronomical telescope with which he can start to make serious observations. This achieved, it is to be hoped that his interest in the work will encourage him to combine his hobbies and make some more advanced telescopes. With this in mind, the concluding chapters indicate some further projects. In general,

I have tried to include information of the sort that I, when first attempting to make a telescope as a schoolboy, would have found most useful. Indeed, I have tried, as far as possible, to remember the problems I then faced, and to offer the advice that I was most in need of, but could not find. I believe that I would, as a young beginner in telescope-making, have valued this book; and if this is so, then others will find it useful today.

James Muirden

CHAPTER ONE

Astronomers and their Telescopes

You can be an astronomer without owning a telescope at all. Even today, there are some observations that are best made with the naked eye. In the early days of civilisation, long before telescopes or even lenses were thought of, astronomers were scanning the heavens with nothing more elaborate than simple sighting instruments; and these observations enabled them to predict eclipses of the sun and moon and to make reasonably accurate deductions about the size of the earth. The greatest of the pre-telescopic observers, the Danish nobleman Tycho Brahe (1546–1601), was able to take sightings on the stars that were, in some cases, accurate to 30 seconds of arc (written 30″), which corresponds to the angle subtended by a penny piece at a distance of 140 metres!

But to find out anything at all about the heavens in close-up, such as seeing the disc of a planet, or observing the faint members of a star cluster, a telescope is essential. The eye is a wide-field, low-power instrument, designed to help us in our normal daily routines; but it cannot perform extreme tasks. If we want to examine a very small object, and hold it just a few centimetres away, it cannot be focused without strain, and a magnifying glass or microscope must be used. At the other end of the scale, its astronomical potential is limited by its low resolving power and its limited light grasp. The purpose of an astronomical telescope is to remedy these defects.

Light grasp

The eye's focusing lens, or *cornea,* carries a diaphragm known as the *iris*. This iris is able to adjust itself spontaneously according to the brightness of the scene, so that the intensity of the image formed on the sensitive *retina* remains more or less the same. In bright daylight, the diameter of the iris is about 2mm, but in twilight or dark conditions it expands to its maximum aperture of about 8mm, and this is the effective aperture of our own personal 'telescope'. It follows from this that if we wish to observe stars that are twice as faint as the naked-eye limit, a telescope with an aperture of twice the area of the fully-expanded iris is needed. Since the area of a circle increases as the square of its diameter, it follows that an 11.5mm aperture telescope would be required. An aperture of 30mm, which is to be found in many binoculars and small hand telescopes, will gather about 14 times as much light as the naked eye, revealing far more *extra* stars than the naked eye can itself perceive on a clear night.

Resolving power

While light grasp indicates the brightness of an object as seen through the telescope, resolving power is an indication of the telescope's ability to show fine detail. Examples of this are markings on the surface of the moon, or on the planets Mars and Jupiter; but it applies equally to the ability of the instrument to divide, or resolve, a pair of stars that lie very close together in the sky. These so-called *double stars* form very convenient tests of a telescope's resolution.

One's first reaction might well be that resolving power depends on magnification, and that what cannot be seen with a low power will be revealed with a high one. This would indeed be so if a point object (such as a star) gave a point image of no area, but the wave nature of light does not permit a point image to be formed in a telescope. The light waves, as they converge to a focus, interfere with each other and produce a characteristic disc and rings pattern, as shown in Fig. 1. A star, when viewed through a perfect telescope, appears not as a point but as a diffraction disc (known as the *Airy disc* after the Astronomer Royal, Sir G. B. Airy (1801–92), who examined its nature), surrounded by two or three

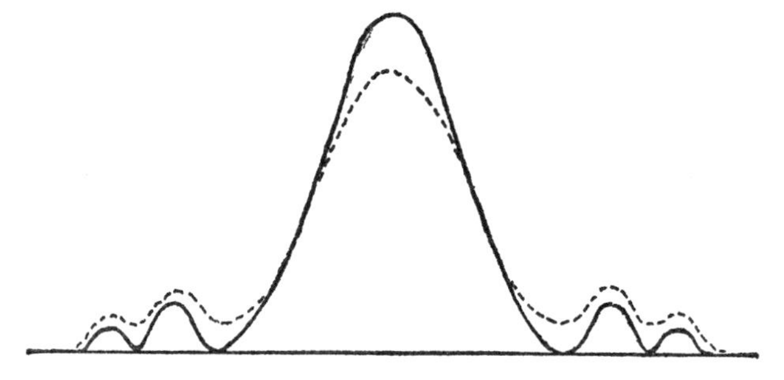

Fig. 1 The solid line represents the light distribution in a perfect star image: the Airy disc and rings. If the optics are imperfect (dotted line), the central disc is dimmed and the rings are brightened, producing a loss of contrast

very faint diffraction rings. The chief difference between a good telescope and a poor one lies in the quality of the image; if the optics are bad, a star will appear as a fuzzy blur with no trace of disc or rings.

The diameter of the diffraction disc varies inversely with the aperture of the telescope. An instrument of 30mm aperture should show a star as a disc about 4″ across. Since the diffraction disc is the building block with which the telescope makes up its images, we say that the resolving power of a 30mm telescope is 4″, and it is useless trying to separate a double star whose components are only 2″ apart, no matter how high a magnification is used.

Development of the telescope

The development of the telescope has been the prerogative of astronomers and astronomically-minded opticians and engineers, and it is easy to understand the romance of struggling, with glass and metal, to create instruments that can carry the human eye and brain far out into space, back millions of years in time, to observe the profundity of the universe. The inspiration, down 350 years of effort to improve our view of the cosmos, can be summed up by the plea for 'More light!' that was voiced by George Ellery Hale (1868–1938), the founder of three of the greatest observatories in the United

States. The great new telescopes of history have usually been born at critical times in the development of the science, when observations at the limit of existing equipment had suggested such intriguing possibilities that the problems faced in the building of larger instruments were somehow overcome. The situation today is not so much that of wanting still larger telescopes than are already in existence, but of wanting more of them, and in better observing sites. It is generally agreed that there is little point in trying to build huge new telescopes when they must operate at the bottom of the earth's dirty and turbulent atmosphere; the new hope is that relatively modest equipment, carried aloft to form orbiting space observatories, will take the next big step in man's probing into deep space.

In retrospect, it seems extraordinary that the telescope was not invented centuries before the records suggest, for glass lenses were known in the 13th century as an aid to defective vision. Indeed, some extracts from the writings of Roger Bacon (1210–94) indicate that he might have developed a crude form of telescope, and there are other vague references to the possibility in 16th century records. But the 'official' birthdate of the telescope is 1608, when a Dutch spectacle maker named Hans Lippershey applied for a patent; it is significant that he failed in his attempt, because other claimants to the invention were already coming forward. In the following year the news came to the philosopher Galileo Galilei (1564–1642) at Padua, in Italy. He was one of the very first people to turn the telescope to the sky, and this momentous event turned Galileo into an astronomer, and turned astronomy into a science of discovery.

Galileo's telescope

This telescope, like Lippershey's, was of the *refracting* form, which means that a lens or *object glass* is used to form an image of the object to which it is pointed. A lens capable of doing this is said to be *positive*, because the light passing through it is refracted inwards, in a converging cone, to pro-

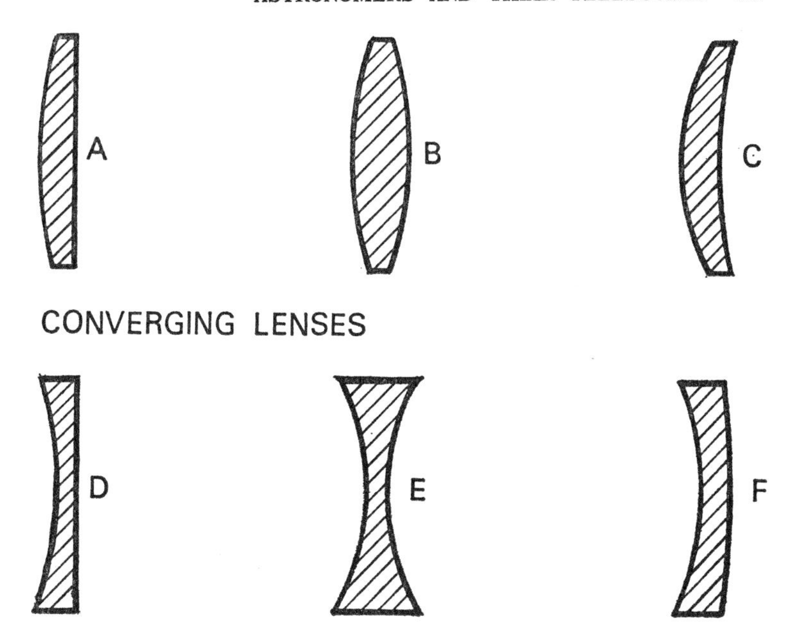

Fig. 2 Different lens shapes. A: plano-convex. B: biconvex. C: convex meniscus. D: plano-concave. E: biconcave. F: concave meniscus

duce a real image that can be caught on a screen or examined with an eyepiece. Lenses are classified according to their cross-sectional shape (Fig. 2), and their surfaces can be convex, concave, or flat, in various combinations. However, positive lenses are always thicker at the middle than at the edge, the reverse being true of *negative* or diverging lenses, which cannot form a real image because the light rays are refracted outwards rather than inwards.

The simple telescope with which Galileo made his great

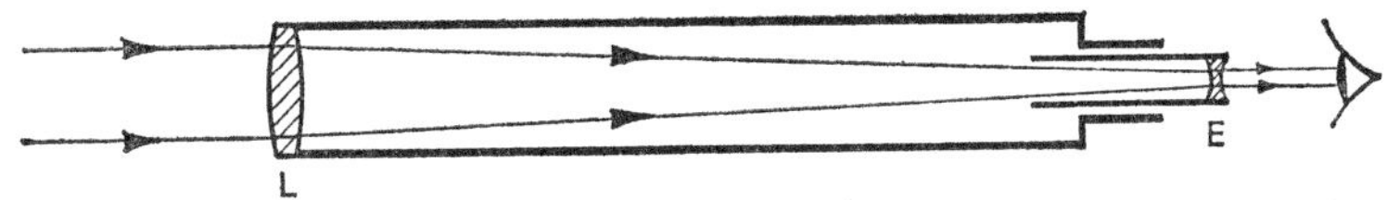

Fig. 3 Galileo's telescope. L: object glass. E: eyepiece. The drawing is not to scale

discoveries is shown in schematic form in Fig. 3. His object glass was a biconvex lens about 30mm across, and it formed the image of a star or other distant object about 120cm down the tube. This distance is known as the *focal length* of the lens. To examine this image, a powerful lens of short focal length (the eyepiece) was made a sliding fit into the lower end of the tube for focusing purposes. Galileo used a negative lens as his eyepiece, but these are rarely used today, as the field of view obtained with such a lens is very small. Galileo's eyepiece had a focal length of about 40mm. The magnification of the telescope, ×30, is equal to the focal length of the object glass divided by that of the eyepiece, a rule that holds good for any type of telescope. It follows that different magnifications can be obtained by using eyepieces of different focal length.

We have here the rudiments of a telescope: an object glass to collect and focus light, and an eyepiece with which to examine the image. Galileo achieved the simplest possible instrument, which could be duplicated by anyone today for £1 or so by purchasing two lenses and mounting them in a cardboard tube, as described in the next chapter. What did he achieve with the 'old discoverer', as he fondly called it? In summary, it made more sensational discoveries than any other scientific instrument. He saw that the planets, the 'wandering stars', showed discs and must be other worlds. Venus passed through phases, and Jupiter had four satellites revolving around it. Then moon was covered with mountainous detail, and the sun revealed temporary dark spots on its surface. Finally, the naked-eye stellar limit was left far behind as the telescope brought into view a whole new realm of faint stars.

The problem of chromatic aberration

More great discoveries followed during the 17th century, all made with enlarged versions of Galileo's refracting telescope. However, astronomers quickly realised that large telescopes, giving brighter images and capable of higher magnification, could not simply be produced by scaling up Galileo's suc-

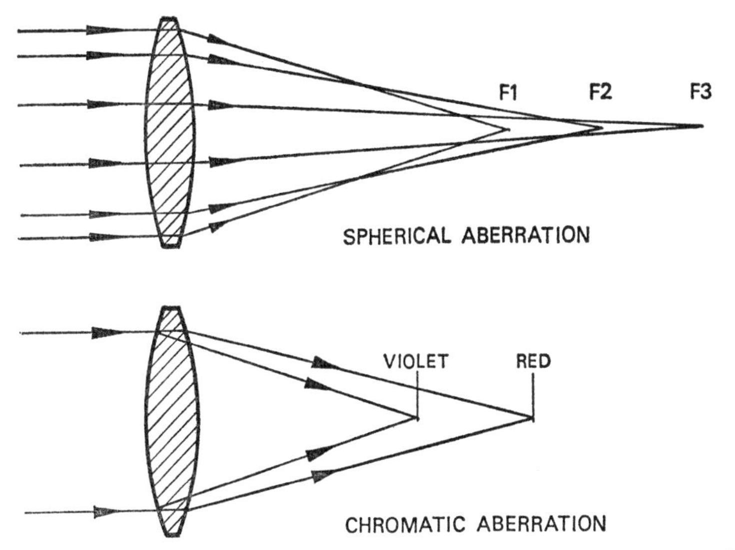

Fig. 4 Spherical and chromatic aberration. In a lens or mirror with spherical surfaces, rays from a distant object striking the margin are brought to a closer focus (F1) than those near the centre (F3). Rays between these come to an intermediate focus (F2). The effects in both diagrams are exaggerated

cessful design. If they did, two unwelcome effects conspired to destroy the sharpness of the image. The first was *spherical aberration*, or the failure of rays passing through different zones of the lens to come to a common focus (Fig. 4). The second, *chromatic aberration*, came through the tendency of the lens to divide or *disperse* the light from the object into its component colours, so that a series of different coloured images was formed, the violet image being nearest the lens and the red image furthest away. In other words, the astronomer could observe a red star with a blue halo, or a blue star with a red halo, according to where he focused his eyepiece – but he could never see a white star with no halo at all!

Of the two effects, chromatic aberration was by far the more serious; in fact, it was so severe that spherical aberration probably went unnoticed. Opticians found, however, that they could minimise the effect by making the focal length of the lens very long in proportion to the aperture, stretching the colours out so that there was less mutual interference. The ratio, the *focal ratio*, is a very important factor in telescope design. Galieo's object glass had a focal ratio of 120cm/3cm or 40, usually written f/40. This was reasonably satisfactory for an object glass of only 30mm aperture; but if the aperture was doubled, to 60mm, an f/40 lens gave an image that was twice as coloured. So a 60mm lens had to be ground to f/80 for equivalent performance, and enormously long telescopes resulted. Some famous instruments of the period included the 50-metre telescope of Johannes Hevelius of Danzig (1611–87), who published the first detailed map of the moon, in 1647; the 7-metre of Christian Huyghens (1629–95), a Dutch astronomer-optician who discovered the rings of Saturn in 1657; and the series of long telescopes of up to 41 metres focal length used by J. D. Cassini (1625–1712) at the Observatory of Paris between 1671 and 1684, with which he discovered four of Saturn's satellites. None of these lenses were more than 20cm or so in aperture; many amateurs today own larger-aperture instruments that are not more than 1½ metres long.

The reflecting telescope

The unwieldy nature of these 'aerial' telescopes, which had to be supported by masts and rigging, made astronomers seek more compact solutions. Out of this arose the *reflecting* telescope, which uses a concave mirror rather than a positive lens to form its image. In the system proposed by Sir Isaac Newton (1643–1727), who himself made the world's first reflecting telescope, in 1671, a concave mirror was placed facing upwards at the bottom of a tube whose top was left open (Fig. 5). The convergent rays returned up the tube, and were reflected through a hole cut in the side by a small flat mirror, which obstructed only a small percentage of the incoming light. A mirror has the great advantage over a lens that it reflects light of all colours evenly, and the effect of chromatic aberration does not arise, so that short focal ratios can be

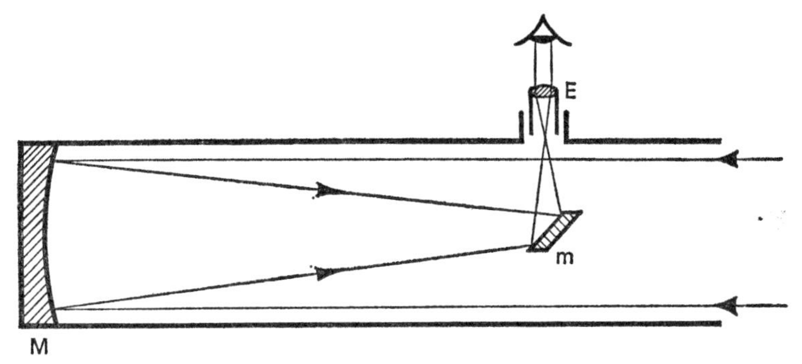

Fig. 5 Newton's telescope. M: main mirror. m: plane mirror. E: eyepiece

used. On the other hand, mirrors of short relative focal length suffer seriously from spherical aberration, and it was many years before opticians learned how to produce the right curve or *figure* on their mirrors to alleviate the defect.

Spherical aberration

The problem of spherical aberration, which will concern us through much of this book, is illustrated in Fig. 6. The rays

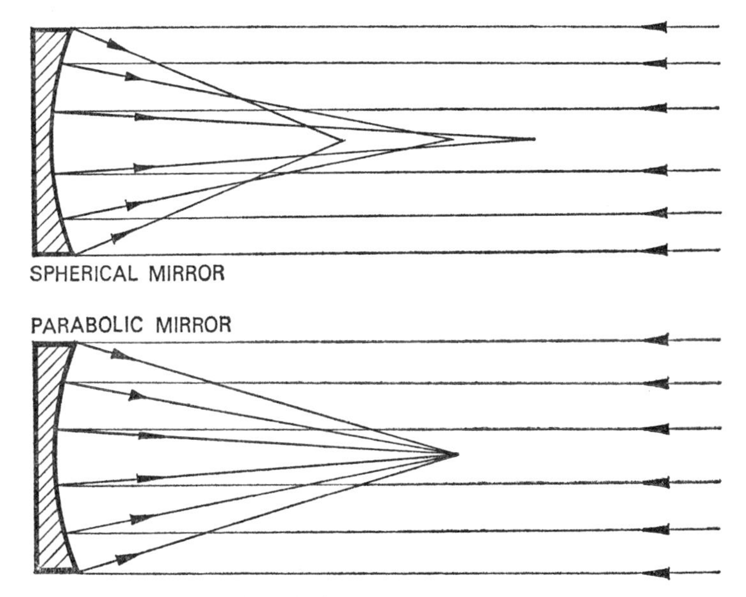

Fig. 6 Correcting the spherical aberration of a mirror

of light from a distant object, such as a star, can be considered as travelling in parallel lines, and astronomical telescopes are concerned exclusively with focusing 'parallel' light. Upon striking a spherical mirror, or a lens with spherical surfaces, it is found that the edge or marginal rays are brought to a shorter focus than the central or paraxial rays, so that nowhere is a sharp image of the object to be found. The distance between the marginal and paraxial foci is an indication of the amount of spherical aberration present, and, for a given focal length, it increases as the aperture increases. By making the focal ratio large, spherical aberration can be reduced to an insignificant level; but if a compact telescope is required, the problem must be overcome by figuring the surface to a non-spherical shape. In the case of a concave mirror, this shape is a *parabola*, which has a slightly shorter focal length at the centre than at the margin. The parabolic shape exactly balances the spherical aberration, allowing all the rays reflected by the mirror to meet at a sharp focus.

The difference between spherical and parabolic surfaces on the typical astronomical mirror is very small, and can be measured only by optical tests. The mirror described later in this book, with a diameter of 15cm and a focal length of 150cm (f/10), requires its centre to be deepened by only 0.3 of a wavelength of yellow light (or about 0.00015mm) to produce a paraboloid from a sphere. Even a large observatory reflector, with an aperture of, say, 50cm, and a focal length of 250cm (f/5), requires only 8 wavelengths of glass (0.004mm) to be removed from its surface. It can readily be seen that the working of such minute amounts of material is a most delicate operation.

Sir William Herschel's telescopes

Such skills were not available in the 17th century, when the Newtonian telescope was introduced, although even the crude optical methods available then produced reflecting telescopes that made a great impression when compared with the spindly refractors of the time. John Hadley (1682–1743), in 1721, presented to the Royal Society a 15cm f/10 Newtonian reflector that compared favourably with a 37-metre aerial telescope used by Huyghens! But the reflector did not achieve its true potential in the field of observation until the Hanoverian musician William Herschel (1738–1822) became fascinated by astronomy and decided to make his own instruments.

The decision was forced upon him through the lack of available equipment, and between the years 1773 and 1789 he produced a succession of reflecting telescopes of progressively greater aperture, culminating in the famous 'forty-foot' reflector of 120cm aperture (48 inch), working at f/10. In both size and optical quality, Herschel's instruments surpassed all that had gone before, and the work he achieved with them was prodigious. Those were the days before the discovery that glass could be coated chemically with silver to make it reflective, and most mirrors were made of a bright compound of copper and tin known as 'speculum metal', which was brittle and hard to work and reflected only about 60 per cent of the light falling on it, compared with over 90 per cent for fresh silver. The drawbacks of having to use this substance caused the reflecting telescope to suffer a decline after Herschel's death, until glass mirrors, coated on their front surface with silver, began to appear some thirty years later.

Herschel, like his contemporaries, was handicapped by the lack of a sensitive test for the optical shape of his mirrors. It was therefore necessary to test each mirror, during the polishing process, on a star or some distant daylight object; by placing masks over the mirror to uncover different zones, Herschel was able to deduce the focal length of each zone and to polish down the high regions accordingly until they all came to a common focus. The inconvenience of such a procedure does not need to be stressed, and it is probably significant that many reflecting telescopes of the period were of long focal length, often f/10 or more, which allowed approximately spherical surfaces to give acceptable results. There is no doubt at all that some of Herschel's mirrors were of superb quality, and it is probably true to say that he made more telescopic discoveries than any other observer, before or since.

Achromatic refracting telescopes

The refracting telescope, meanwhile, had undergone a transformation when the London optician John Dollond (1706–61), round about 1758, began manufacturing *achromatic* ('colour-free') refracting telescopes. It had been discovered that combining two lenses made of different kinds of glass caused the chromatic aberration of one to be cancelled out by that of the other (Fig. 7). The types of glass chosen, which are still used today in achromatic combinations, are known as *crown* and *flint*, crown glass being harder, lighter in weight, and of weaker dispersive power than flint. Although the first achromats were small affairs, or perhaps 7cm aperture, improved

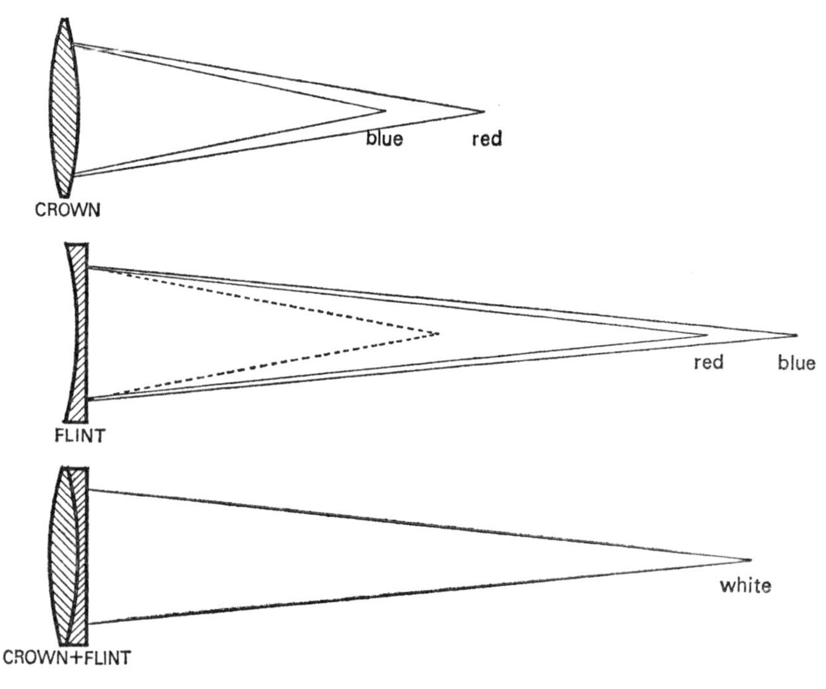

Fig. 7 The principle of the achromatic lens. The dispersive tendencies of the crown and flint lenses work in the opposite direction, and tend to cancel each other out

methods of optical glass manufacture produced ever-larger discs. However, it was still not possible to make refractors of very short focal ratio, and f/15 to f/20 was and still is the usual design, the reason being that some slight secondary colour is still visible even in a nominally 'achromatic' lens.

One of the best-known of the early achromatic refractors was the 24cm (9½inch) which was installed at the Dorpat Observatory (now in Estonia) in 1825. Made by the brilliant German optician Joseph Fraunhöfer (1787–1826), it was used by Wilhelm Struve (1793–1864) to examine about 120,000 stars; and his measurements of over 3,000 double stars stand as a tribute to the excellence of the instrument. For micrometrical work, where stability and constancy of the optical components of a telescope are important, the refractor can offer better service than the reflector, whose mirrors are difficult to hold rigidly in their cells and whose surfaces require periodic re-coating because of tarnishing.

The modern giants

By the end of the 19th century we find ourselves entering the 'modern' age of telescope-building, and many of the great instruments created at that time are still in constant use. In 1897 the greatest refracting telescope ever built, with an aperture of 100cm (40 inches), commenced work at the Yerkes Observatory, Wisconsin. This was the first of four epoch-making telescopes installed through the energy of Hale in his quest for 'more light'. He realised, however, that in the Yerkes refractor he had reached the practical limit of this type of instrument; not only was its tube almost as long as a cricket pitch, requiring a huge and costly dome, but a larger object glass would have been difficult to grind and mount because of the flexibility of the relatively thin discs of glass. So Hale turned to the reflector, and his efforts produced the 1½- and 2-metre (60- and 100-inch) telescopes at the Mount Wilson Observatory in 1908 and 1917 respectively, and the posthumously-completed 5-metre (200-inch) at the Mount Palomar Observatory, which went into service in 1948 and is still the largest optical telescope in the world. More recently, the ranks of the giants have been reinforced by the 3-metre (120-inch) and 4-metre reflectors at the Lick and Kitt Peak Observatories, also in the United States, while a 3.8-metre (150-inch) telescope is being installed at the Anglo-Australian Observatory at Siding Spring, Australia. A 5-metre reflector being built by Soviet astronomers has yet to be completed.

These huge telescopes are rarely, if ever, used visually. The objects they study are usually faint, and the photographic emulsion is much more sensitive to radiation than is the eye; so in a sense it is more correct to think of them as giant cameras. Exposures may last for several hours, and to counteract the earth's daily rotation, which makes celestial objects appear to move across the sky from east to west, these telescopes are mounted *equatorially,* on an axis that is parallel to that of the earth. By driving the telescope around this *polar axis* so that it turns once a day in a sense opposite to that of the earth's spin, the telescope remains pointing to the same region in space. All professional telescopes, and many amateur instruments, are equatorially mounted, since even for visual work automatic following of an object is a great convenience.

Compound telescopes

Despite their colossal apertures, large reflecting telescopes are relatively compact. For example, the Mount Palomar instrument has an f/3.3 mirror, or a focal length of about 17 metres, which is shorter than that of the much smaller Yerkes refractor. Such 'fast' mirrors are extremely difficult to figure to a true parabolic curve – it took several years to figure the Palomar mirror – but tube length is a critical factor, since a long telescope is much more difficult and expensive to mount than a short one. Most large reflectors work at f/5 or shorter. However, for some observational work a long focal length is necessary, and this effect can be achieved by placing a relatively small convex mirror inside the tube near the top end to reflect the light from the primary mirror back down the

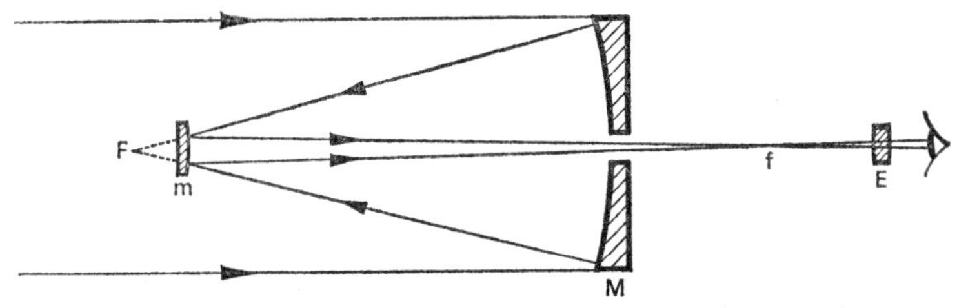

Fig. 8 The Cassegrain telescope. M: main or primary mirror. m: secondary mirror. F: primary focus. f: secondary focus. E: eyepiece

tube before it has come to a focus. Such an arrangement, one of many a *compound* telescope can take, is called a *Cassegrain* telescope after the Frenchman who proposed it in 1672. It is represented in Fig. 8. It will be seen that the steep cone of light striking the small or *secondary* mirror is replaced by a narrow one, which usually is allowed to pass through a small hole cut in the centre of the primary mirror, behind which the Cassegrain focus is located. The *effective* focal length of a Cassegrain telescope is equal to the distance the narrow cone would have to be extended until it equalled the diameter of the primary mirror. Most Cassegrains work with an amplification of about 5, so that, for example, the primary focal ratio might be 4, and the Cassegrain focal ratio 20. It is clearly a great saving in weight and expense to be able to contain an f/20 focal length within the extent of an f/4 tube! All large reflectors are of the Cassegrain form, with probably several secondary mirrors to give different amplifications available, while prime-focus photography is also possible by placing the sensitive emulsion inside the tube, at the direct focus of the primary mirror.

In the present century, the astronomer's instrumental armoury has been extended by new optical designs. The purpose of this introductory chapter has, however, been to give the reader some initial insight into the kinds of astronomical telescope he is likely to use or construct, and the choice generally lies between refractors and Newtonian reflectors. The possibilities offered by more advanced optical systems are outlined in Chapter Ten.

CHAPTER TWO

Making a simple Refracting Telescope

It is perfectly possible for the beginner, intrigued at having read an astronomical book and wishing to see the heavens in close-up, to go straight to a camera or optical shop and buy a cheap refracting telescope. But, unless he is prepared to spend £20 or more, he is likely to be disappointed once the excitement of the first views has passed. Indeed, some telescopes costing three times as much can be very poor buys indeed. A recent test report on telescopes in the Junior Astronomical Society's magazine *Hermes* (Vol. 20, page 45, 1973) was critical of a 76mm refractor costing over £70 and widely marketed. It commented: 'Chromatic aberration noticeable on all powers. Bright star image circular with several diffraction rings, but with false colour . . . Play in equatorial head made image very unsteady.'

The beginner will learn far more about telescopes, and will save a great deal of money, by making his first telescope himself. It need not take more than a week of evenings to produce a simple refractor capable of showing lunar detail, Jupiter's disc and satellites, and possibly the rings of Saturn, at a cost of less than £5. It will be a telescope giving at least as satisfactory performance as the one Galileo used to launch the modern astronomical age. Most important of all, the views it gives will be enhanced by the observer's knowledge that he created the instrument himself! In this small way will be bestowed the impetus to tackle the far more ambitious instruments described in the following chapters of this book.

The instrument will consist, as do all refracting telescopes, of the following basic parts.

The object glass

A positive lens with an aperture of about 40mm and a focal length of about 1 metre is required. Optically, the longer the focal length the better, since spherical and chromatic aberration both become less severe as the focal ratio increases, but the unwieldiness of a very long tube sets a practical limit. In use, the lens is stopped down to about 35mm (f/30), to give reasonable performance; indeed, it may be found that a 30mm stop, although reducing the brightness of the image by a significant amount, gives an improved view through the extra whiteness and sharpness of the details. It is in such matters that trial and error play their part, and the constructor learns from his experiments until he has achieved the best compromise. Suitable lenses can be obtained from the suppliers listed in Appendix A. Alternatively, use can be made of a camera supplementary lens, designed to be fitted over a standard camera lens to allow it to focus for close-up work. These come in various standard diameters. A 1-dioptre lens has the required focal length of 1 metre (the term *dioptre* indicates the power of a lens, or the inverse focal length in metres, and is used in ophthalmic work for rating spectacles; a 2-dioptre lens has a focal length of 50cm, a 3-dioptre lens 33.3cm, and so on). Supplementary lenses are usually of the meniscus form, as opposed to the more common plano-convex type; in both cases, however, the lens gives the least amount of spherical aberration if the convex side is turned to face the object.

The eyepiece

Single-lens eyepieces, of the type used in early telescopes, suffer from a number of disadvantages, notably chromatic aberration and a small field of view. They have been entirely replaced today by multi-lens eyepieces, which remedy these defects and others in varying degrees. The commonest, and cheapest, is the *Huyghenian,* which consists of two plano-convex lenses (see page 134), and it is an advantage of a long-focus refractor that a more complicated and expensive eye-piece is not needed, and may in fact give inferior results. An ex-government Huyghenian, the most expensive item in the whole telescope, will probably cost £2–3.

Eyepiece focal lengths range from about 4mm to 40mm or more, and the choice naturally depends on the magnification required. The lower limit can be set at once by considering the aperture of the human pupil. The cylinder of light entering the object glass of a telescope emerges from the eye-piece minified by a factor corresponding to the magnification of the telescope. For example, a 50mm telescope magnifying 10 times would pass into the eye a beam of light (the *eye-beam*) only 5mm across; if the power were increased to 25 times, the eye-beam would be 2mm across. Bearing in mind that the greatest opening of the pupil (which occurs in dark conditions) is about 8mm, we see that a power giving an eye-beam greater than this diameter would not be passing all the light focused by the object glass into the eye, and the aperture of the telescope would effectively be stopped down by the pupil*. A magnification of 6¼ times would, therefore, be the lowest usable power with a 50mm telescope; but such a power would be far too low to reveal all the detail that even a non-achromatic lens is capable of defining. On the other hand, high magnifications are limited by the quality of the image formed by the objective, the difficulty of holding the tube steady, and the diminished field of view.

Field of view

This consideration is most important. If an eyepiece is held up to the bright sky with the eye in the normal viewing position, a circle of light is seen whose diameter may be anything from 25° to 60° or even more. One way of judging this diameter is to view, with the other eye, the out-

*The minimum useful magnification is therefore given by 10/8=1.25 times the aperture in centimetres, or 3.3 times the aperture in inches.

stretched hand at arm's length, and to see how many multiples of the distance from the thumb to the little finger can be fitted into the circle: each hand's spread measures about 20°. The typical Huyghenian eyepiece has an *apparent field of view* of about 40°. To find the *real* field of view afforded by the telescope, the apparent field is divided by the magnification given by the eyepiece; hence, in this particular case, ×40 will show a 1° circle of sky, while ×80 will show only ½°, which corresponds to the diameter of the sun or moon. This is not a very large amount of sky, as will be appreciated when trying to coax the moon, or a planet, into the field of view of a powerful eyepiece! A magnification of about ×40 will probably be found to offer the best compromise. A 25mm eyepiece is required for such a power.

The telescope tube

The main purpose of a telescope tube is to hold the optical components (in this case, lens and eyepiece) in their correct relative positions. Several precautions must, however, be observed.

(a) The object glass must be mounted with its optical axis (the line passing at right angles through its centre) coincident with the axis of the tube. If it is out of square, it will not pass through the centre of the eyepiece and star images may appear as V-shaped blurs (*coma*), or even as lines or crosses (*astigmatism*) if the maladjustment is gross. On the other hand, its retaining rings or cell must not exert any pressure on the edge of the lens, or the glass may be twisted enough to distort the image.

(b) The eyepiece must also be well aligned with the tube and the optical axis. The image is focused by fitting the eyepiece into a short tube (the *drawtube*) that slides smoothly inside the main tube.

(c) When observing an object that is very near a bright source of light, such as a street lamp or even the moon, reflections of this light inside the tube can brighten the field of view and fog the image. These reflections can be reduced somewhat by painting the interior of the tube matt black, but a couple of diaphragms or *stops* inside the tube will almost eliminate the effect.

(d) On damp nights, dew will form on cold surfaces exposed to the air; if the object glass dews, definition will be lost. Mounting the lens a distance of 3 or 4 diameters down from the top of the tube, so forming a *dewcap*, will considerably delay the onset of dewing, and will also help to remove the reflection effect mentioned above. Alternatively, an extension can be added to the existing tube.

The tube can be made of metal, fibreglass, PVC, or even cardboard. The lighter it is, the easier it is to mount satisfactorily.

The mounting

Although no telescope can perform better than its optical specification will allow, it will fail even to approach this limit if its mounting does not allow the tube to be pointed easily at any desired object, and, once pointed, holds it steady. It is the writer's experience that at least as many telescopes suffer from poor mountings as from defective optics; all too often the beginner, having assembled his telescope, is so anxious to use it that he rests it on a wall, or against a convenient post, and hopes to see details on a planet as it dances in the field of view. Not surprisingly he is disappointed, and abandons the instrument as worthless before it has been given a chance of showing its potential.

Most small refractors are mounted on a pivot, or in a fork, located somewhere near the centre of gravity of the tube. This bearing is made sufficiently tight to hold the tube in the desired position, but loose enough to allow controlled movement. However, such an arrangement is most unsound, for the tube is free to vibrate along its length at the slightest touch or puff of wind. A far better method is to hold the tube at two points along its length, for in this way vibration is quickly damped out.

A simple and effective mount for a small refractor is shown,

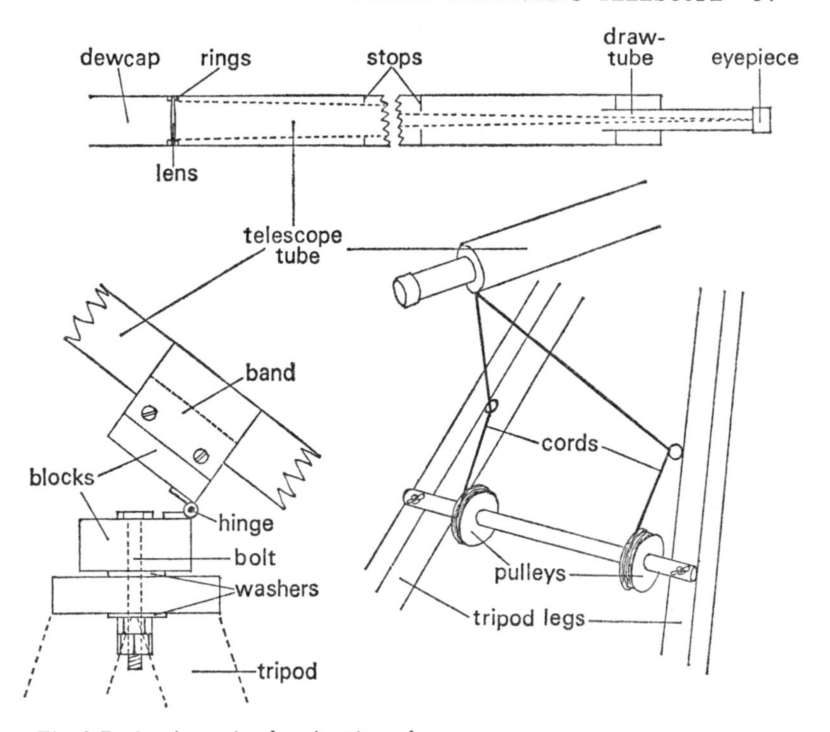

Fig. 9 Design for a simple refracting telescope

together with other details of its construction, in Fig. 9. A tall tripod, about 2 metres in height, has at its top a vertical pivot carrying a block. The tube is attached to this block, using a horizontal hinge, at a position about a third of the way down its length. At the bottom of the tube are attached two cords which wind up on pulleys fixed to two of the tripod legs. The top of the tube is weighted, so as to keep the cords pulled tight at all altitudes. By adjusting the lengths of the cords the tube can be made to move, under fine control, in altitude or azimuth or in any combination of the two. For large changes of azimuth, the whole tripod is moved round.

Construction of the telescope

The important details are shown in Fig. 9, but the maker will adapt them to his own requirements, as we are concerned with principles rather than practice. The very first thing to do is to measure the focal length of the lens, by imaging the sun or moon on a piece of card, rather than accepting the supplier's word for it! Ideally, the tube should be large enough to accept the lens as an easy fit, but if it is considerably too wide, the point at which the lens is mounted can be reduced in diameter by a liner, or the lens can be mounted in a tube of the correct diameter which is then fitted inside the main tube. Two rings retain the lens in position, but they are not finally fixed until the lens has been squared-on by examining the image of a star.

The tube is cut off at the lower end so that the focus falls about 10cm beyond it, and a tube about 20cm long, which slides smoothly inside the main tube, carries the eyepiece. This can be pushed into the end of the drawtube, or fastened with adhesive tape. Proper two-piece drawtubes, with standard eyepiece fittings, can be purchased from telescope suppliers, but will not necessarily give better results than a simpler arrangement. Two cardboard stops are fitted as shown after the interior of the tube has been painted dull black. This completes the construction of the telescope itself.

The tripod can be made from 3 2-metre lengths of 75 x 25mm deal, attached by hinges to a hexagonal top. The legs then fold together for storage. Three stretchers, attached half-way down the legs by screws and wing nuts for quick removal, aid stability. Take a piece of hardwood about 10cm square and 5cm thick, and attach a similar block to it with a heavy-duty hinge as shown. This second block has a V-shaped groove cut in it to act as a rest for the telescope tube, which is secured to it by one or two metal bands. The best position for the fixing is determined by experiment. The first block is then secured to the top of the tripod using a central bolt, with a large washer between the two surfaces allowing smooth rotation.

A piece of metal, sufficiently heavy to give the tube a definite downwards bias even at high altitudes, is secured near the object glass. Two cords now pass from the eyepiece end of the tube to the stand. A very simple, and surprisingly effective, way of making these adjustable in length is to pass them through two eyelets on the legs and then down for two or three turns around one of the stretchers, which is of circular cross-section. A small weight at the end of each cord holds it in position and maintains friction, and the observer can slowly twist the turns on the wood with his fingers. The job will, however, be more workman-like if two stiff rollers are threaded on the stretcher and the cords wound up on them.

This cord system is a simple version of the telescopic rods often seen on old old-fashioned brass refractors, and will work perfectly well if the telescope's altitude does not exceed about 70° – at which point the instrument will, in any case, be most uncomfortable to use because of the severe neck-twisting involved.

Of course, it is possible to make a more elegant mounting, but a telescope such as this is not intended to last a lifetime and there is not, in the writer's view, much point in spending a great deal of time and money on its construction. Simplicity combined with efficiency is the keynote – as should, indeed, be the case with all instruments. It will have served its purpose if it shows objects distinctly enough to whet the observer's appetite for more.

Adjusting and testing

It is first necessary to square-on the object glass. Bring a bright star into the field of view, and examine its image with the lens at full aperture. Ignoring, for the moment, the colours, push the eyepiece inside and outside the best focus and observe whether or not the expanded images of the star are circular. If they are not, the fault will probably take the form of a flare to one side of the disc (the same side inside and outside focus), while the focused image itself will carry a tail. This is the defect of coma. The object glass can be adjusted by trial and error, but it saves time to remember that the head of the comatic image points towards the optical axis of the lens. In other words, if the tail is to the right in the eyepiece, the true axis lies over to the left, and the right side of the object glass should be pushed away from the observer. If the image is in the form of a line, whose axis moves through a right angle as the position of best focus is passed, then astigmatism is present. If it cannot be relieved by adjustment, it is either inherent in the lens, due to poor manufacture, or the lens is being pinched in its cell.

We can now turn our attention to the colours produced by the object glass. We have already seen (Fig. 8), that a single lens focuses blue light short and red light long, with yellow light – the colour to which the eye is most sensitive – falling somewhere between the two. Selecting first of all the best focus, a white star will appear as a yellowish point surrounded by a haze formed from the other out-of-focus colours – the outer halo is reddish-purple, but inside this we encounter the intermediate tints of orange and green. Pushing the eyepiece towards the object glass, the 'star' turns green and finally blue, and the red outer halo expands because we are moving still further inside its point of focus. Taking the eyepiece back, the red focus is reached. Note how the blue halo now formed is much less conspicuous than was the red one at the blue focus; this is because the lens spreads the blue light much more than the red, giving less concentration of colour.

We would not mind very much a star appearing yellow or green instead of white, if it were not for the out-of-focus colours interfering. Stopping down the lens helps to achieve this by narrowing the beam and reducing the intensity of the out-of-focus colours by a greater proportion than it reduces that of the focused image. In theory at least, since the focused image of a star is a point and the halo is an area, the latter is dimmed as the square of the former, so that reducing the aperture by half will make the halo only a quarter as obvious. Experiments with different sizes of stops before the object glass will indicate the best compromise between image and halo intensities.

Another way of achieving this colour selectivity is with filters. The best colour to choose is red, since the light is least dispersed, and the lens can now, of course, be used at full aperture. A deep red gelatine filter, of the kind used in colour-separation work, cut carefully into a circle and set in front of the object glass, will probably give more distinct views of bright objects such as the moon, Venus, and Jupiter, than a stopped-down, non-achromatic system.

Such a telescope will do much to satisfy the first needs of the 'star-struck' enthusiast. He will have saved a lot of money and he will have learned something about optics. The next step is the big one!

CHAPTER THREE

Planning a Reflecting Telescope

With the possession of a 15cm reflecting telescope of good quality, the observer is equipped to undertake most branches of astronomical study with every hope of success. Not the least of the attractions of a telescope of this size lies in the fact that it is small enough to be really portable. Most observers have to endure obstructions in certain parts of the sky, and a large equatorially-mounted instrument, which must be sited permanently and have its polar axis carefully aligned to be of fullest use, will almost inevitably suffer from important 'blind spots'. A movable 15cm telescope will always be of use, and its advantages are not always appreciated until it has been replaced by a larger telescope. If you make a good 15cm reflector, you should keep it, particularly if a fixed instrument is acquired.

Temperature effects

Generally, too, small instruments give more perfect images – though naturally on a smaller scale – than large ones, and are not always out-performed in proportion to their apertures. It must be rare indeed for the owner of a 30cm reflector to see a star showing its theoretical Airy disc and rings, but such an experience is common with a telescope of half the aperture. The increased destruction of the image due to atmospheric turbulence is one part of the story, but at least as important is the sensitivity of a large mirror to the local effects of changing air temperature. This is a most important matter. Glass has a poor coefficient of thermal conductivity, and at nightfall, when observations usually commence, the air temperature is usually dropping much faster than the glass can lose its heat. Therefore we have a situation where the surface of the mirror is warmer than the air, so that the air in contact with the mirror is heated and raises up across the incoming and reflected beams of light. Since the refractive index (see page 144) of air changes with its temperature, the light forming the image is forced to pass through a layer of optically inhomogeneous waves. Not surprisingly, the images produced by a telescope under these circumstances is very far from the ideal 'disc and rings' to be expected on paper!

The seriousness of these 'mirror currents' depends on the temperature difference between the mirror's surface and the night air. On a summer's day, it is quite possible for the temperature of the mirror's surroundings to reach 25°C. for several hours in the afternoon if the instrument is stored in a shed, or under canvas, in full sunlight, whereas the night air temperature may fall to perhaps 15°C. soon after dark. Experiments (*Journal of the British Astronomical Association,* Vol. 82, page 274, 1972) have shown that a glass mirror 30mm thick (somewhat thicker than is usual for a 15cm reflector), can lose its heat sufficiently fast to keep very near to the falling air temperature, whereas the inner parts of a disc 50mm thick – an appropriate thickness for a 30cm mirror – would still be perhaps 2°C. above the ambient temperature after several hours of night-time cooling, and so still releasing weak heat-waves into the air. Perfect definition cannot be expected under such circumstances.

Focal ratio

In support of the 15cm, then, we have the considerable practical advantages of portability, cheapness, relative insensitivity to mirror currents, and simplicity of construction compared with a larger instrument. A further advantage is that the focal ratio can be made larger than normal without involving too cumbersome a tube. A Newtonian mirror should always be made with the longest practicable focal length, and it is perfectly feasible to make a 15cm f/10 telescope, whereas a 30cm instrument cannot be much longer than f/6 without being difficult to mount and awkward to use. It is the writer's experience that a good f/10 15cm mirror gives, in practice, significantly better images than an f/8, besides being easier to make. Long focal ratios are more tolerant of eyepiece defects (particularly achromatism, or the lack of it) and mirror misalignment, and one reason why refractors have a reputation, that is not always justified, of giving 'better definition than reflectors', is the fact that they normally work at about f/15.

Mirror kits

On the other hand, the 15cm f/8 reflector has become so standard an item with many amateurs that ready-to-work kits for this size have been produced. These are excellent value for money (the current price is about £12), since the glass mirror disc and the tool on which it is worked have already been given the correct curve, which requires only a couple of hours' work before it is ready to polish; and they come complete with abrasives, polishing powder, pitch, and other necessary items of equipment that are not easy to come by without finding a friendly optical firm willing to help out. Much time is saved by not having to 'hog out' the rough curve with coarse abrasive, and the writer would certainly recommend these kits unreservedly if it were not that an f/10 mirror is easier to test and may well give better results. In what follows, it will be assumed that an f/10 mirror is being made from the basic materials. The user of an f/8 kit will join the discussion about halfway through Chapter Four. From that point onwards the difference is one of dimensions, and conversions are given where appropriate. Similarly, mounting details differ only in the matter of tube length.

Tube and fittings

As in a refracting telescope, the function of the tube of a reflector is to hold the optical components in their correct mutual positions. The mirror is situated at the lower end of the tube, facing upwards, and there must be adequate provision for squaring-on, so that its optical axis runs accurately up the centre of the tube. On this axis, at the top end of the tube, is a small elliptical plane mirror known as the *flat*, which is inclined at 45° to the axis of the tube and reflects the light through a right angle into the eyepiece. The flat is elliptical so that its outline in the main mirror is circular. A rectangular flat would effectively serve just as well, but since these mirrors are normally purchased ready-made, and good ones are always made elliptical, there is no choice in the matter. This flat can be mounted in a single arm projecting from the inside of the tube, or else sits at the centre of a 4-vaned 'spider'. The focusing tube can be a simple push-pull affair, or have a helical screw, or (best of all) use a rack-and-pinion focusing adjustment.

The simplest possible tube for a reflecting telescope would consist of no more than a single wood plank, about equal in length to the focal length of the mirror, with the mirror cell secured at one end and the diagonal and eyepiece at the other. The drawback of the 'tubeless' telescope is that dew forms very readily on the optical surfaces unless the instrument is mounted in an observatory. It is far better to have a closed or semi-closed tube, and this can be of wood, PVC, fibreglass or metal. (See Fig. 22.)

Altazimuth and equatorial mountings

The mounting for the tube can be either altazimuth or equatorial. Professional telescope–makers have generally helped cultivate the view that *any* astronomical telescope should be equatorially mounted, and so we encounter tiny 60mm refractors mounted on spindly axes with inconvenient counterweights, when the same money could provide a far more rigid and satisfactory altazimuth stand. It is certainly true that the equatorial mounting has the great advantage of requiring movement only around the polar axis in order to keep a celestial object in the field of view, but against this we have the fact that it is much easier to make an effective altazimuth mount since there is less overhang on the axes.

Only for photography, and when the same object is likely to be observed continuously, with high magnifications, do the advantages of the equatorial become in practice decisive; and it should be borne in mind that the alignment of the polar axis with that of the earth needs to be correct to within a few minutes of arc for the utmost efficiency. In other words, a portable instrument is unlikely to find itself returned to the accurate positioning required. Better an altazimuth than a poorly-aligned equatorial; and the latter is a nuisance when trying to find objects near the celestial pole, when large adjustments around the polar axis may be required in order to move the direction of the telescope by only a few degrees. To round out the subject, however, a brief discussion of equatorial mountings appears in Chapter Seven.

A simple mounting

Fig. 23 shows the essentials of an altazimuth stand for a small reflector, combining the desiderata of rigidity, ease of fine adjustment of the direction of the telescope, portability, and simplicity of materials and construction. The basis of the stand is a heavy tripod made of wood, well cross-braced. To the top of this is secured a car's hub axle or a similar heavy bearing, and to this in turn is fitted a triangle which rotates in the horizontal plane. Fine motion in azimuth is obtained by a length of threaded rod attached to the back of the triangle and bearing against a plate fitted to the tripod, the backlash being taken up by a spring; for large movement in azimuth, the whole stand is moved bodily, and the threaded rod is periodically wound back to its starting-point. The telescope tube is attached by hinges to the back of the triangle, and a vertical adjustment rod passes through a locking collar on the front of the triangle, with a screwed rod for fine motion. There is nothing in the entire assembly that cannot be made from standard items, and the most basic workshop will furnish sufficient tools for the work, there being nothing in the nature of turning, milling or precise drilling required.

Such a telescope has been made and tested under working conditions, and the writer has no doubt that it is very near to the best compromise available to the keen amateur wanting to make the best possible telescope in the shortest time – which surely corresponds closely to the wishes of the person reading this book. It is possible to labour for months turning out an instrument that will earn pride of place at a club exhibition, but it is the writer's experience that such telescopes are rarely turned to much practical use. This is in no way intended to decry the spirit of the endeavour, but points up the difference between the hobbies of telescope-making and astronomical observing, which are two very different things! It is a curious and interesting fact that the two rarely go together; the more determined the maker is to produce an elaborately finished instrument, the less likely he is to use it much when it is completed, the next object tending to be the production of a larger and improved model. A functional approach is not only permissible, but desirable, and demonstrates the enthusiasm of the maker to get down to the business of observing. Once the observer has gained experience, and decided on the fields of work on which he wishes to concentrate, he can design himself a more specialised and elaborate instrument.

The next three chapters of this book will be concerned with the production of a 15cm mirror. Chapter Seven will then describe the construction of a tube and mounting along the lines discussed above, although in general detail only, since anyone handy enough to undertake the work will be able to devise his own methods, once he understands the ultimate aim. The optical work is the heart of the matter, and we can begin at once with an outline of the process and a list of essential equipment.

Making the mirror

We begin with two circular discs of glass, one of which is to become the mirror, the other being the 'tool' on which the mirror is ground to curve. The tool is secured to the top of a stand around which the operator can walk, and coarse carborundum powder, mixed with a few drops of water, is sprinkled on its surface. The mirror disc is then ground face-down on the tool, the grinding action causing the mirror disc to turn concave, while the tool becomes correspondingly convex. Once the correct curve has been attained, the coarse abrasive is replaced by a series of finer and finer powders, until the glass surface is so finely smoothed that it is semi-transparent. The mirror disc is then polished by coating the tool with a thin layer of pitch to make a polisher and continuing the same action as before, using optical rouge instead of abrasive. Once the glass is polished, its shape or 'figure' is examined optically and it is worked up to the correct parabolic shape. This figuring process decides whether the finished telescope will define well or badly, and is of critical importance.

Once figured satisfactorily, the concave surface of the mirror is coated with a reflecting layer of silver or aluminium – usually the latter – and it is ready to go into the telescope. It is impossible to give a very reliable estimate of the likely time taken to make a mirror, but grinding and polishing should not take more than 10–15 hours of actual work – this does not include the time taken to make test apparatus, stands, etc. Figuring will probably take some hours. In other words, about a month of normal spare-time work may be needed – but it could be more.

The following list includes all the items that the prospective mirror-maker will need to purchase specially for the job; a list of suppliers of these and other materials is given in Appendix A.

The mirror and tool discs

Glass, on the optical scale, is a floppy substance, and a disc a centimetre thick can be bent by several microns* with the

*One micron measures 0.001mm, and is equivalent to about 2 wavelengths of yellow light. It is easily remembered as 'a millionth of a metre'.

fingers, while if supported unevenly it will sag noticeably under its own weight. Astronomical mirrors must therefore be very thick if they are to preserve their figure, when in their cell, to within a fraction of the wavelength of light, unless an elaborate support system is used. Because of this, a 15cm diameter mirror is usually made of glass about 25mm thick, although the thickness of the tool can be somewhat less, perhaps 20mm.

The material used for the mirror disc can be ordinary thick plate glass; or low-expansion glass, of which Pyrex is a common brand, can be obtained in specially-moulded discs. The optical surface of a mirror made of low-expansion glass is less affected by the warmth of the worker's hands, or the friction of polishing, and so allows testing to take place more quickly after a spell of figuring. However, it is doubtful if there is any observational advantage in a low-expansion mirror over an ordinary plate-glass one until we encounter large and thick discs of perhaps 40cm diameter and over, which are very slow to cool at nightfall and retain thermal bumps for long periods. (As regards the production of mirror currents, there is little to choose between the two materials, their coefficient of thermal conductivity being similar.) In the writer's view, a far more important argument for the use of this type of glass is its greater hardness over plate, with reduced sensitivity to the fine smooth scores known as 'sleeks' which often appear during polishing. This hardness also produces a finer-ground surface at the end of the smoothing process, and a quicker polish.

Unfortunately, many of the low-expansion discs available in this country are marred by a very uneven surface, and extra grinding may be required to get the optical curve down to the bottom of the hollows; worse still, they often contain bubbles that may break through the surface and produce an ugly blemish. Allied to bubbles are swirls or discontinuities in the badly-mixed molten glass, known as 'striae'; if these come through the surface they can leave the effect of a deep and ugly sleek, and nothing can be done to remove it. Bubbles (except perhaps very tiny ones) and inhomogeneity are absent from plate glass because it is produced in a continuous sheet process in which the glass is allowed to cool or 'anneal' from the molten state gradually, so that striae have time to work themselves out; this has to be done because plate glass would shatter if cooled quickly, whereas low-expansion glass is much more durable and does not require any significant annealing. 'Fine-annealed' low-expansion castings can be obtained, but are between two and three times as expensive as the regular kind.

It is for these reasons that the writer would not follow those who recommend low-expansion discs unreservedly. The best such castings available in this country are the Duran 50 discs made by Messrs Schott of West Germany, which, although expensive, are all fine-annealed and practically bubble-free. Castings in this glass are widely used in medium-sized observatory instruments.

On the other hand, if plate glass is used the discs have to be cut to order from the sheet, and delays of several weeks are common unless an optical firm happens to have some in stock. The edges should be ground smooth by the supplier, and not left as nipped from the sheet, for the flaked margins will tend to chip and may retain contaminating abrasives.

Being relatively soft, plate glass is preferable for the tool disc since the abrasive sinks into it and gives a finer finish to the mirror. Both mirror and tool discs should be of the same diameter – preferably slightly larger than the required aperture, to allow for a bevel of about 2mm around the face of the finished work.

Abrasives

The initial object of grinding is to produce a concave curve of the required depth on the mirror's surface. For this, we use a coarse abrasive which will do the work as rapidly as possible. This done, the roughness must be removed by using smoothing powders. We therefore have the two processes of rough grinding and smoothing.

For rough grinding, the abrasive universally used is carborundum (silicon carbide). Carborundum powder has sharp-edged particles that dig deeply into the glass, achieving rapid wear with deep pits. It is graded according to the mesh size (in meshes per inch) of the sieve used to sort out the grains; grade 80 is commonly used for roughing out a mirror's curve.

Carborundum powder, because of its deep action, is not normally used throughout the smoothing process, because finer-acting powders are available; but it may well be used for the initial smoothing. To remove grade 80 pits grade 120 can be used, followed by grades 220 and 320. Following these, the softer-acting aluminium oxide abrasives are recommended; they come under various names, such as emery, alundum, and corundum, but the most widely-available form in this country is aloxite, made by the Carborundum Co. of Manchester. This is graded in a different way, by taking the mean size of the particles in terms of 0.1 micron (0.0001mm). An equivalent to grade 320 carborundum, for instance, would be about 360 aloxite; but such equivalences are misleading, for the latter, because of the more rounded shape of the grains, will produce a finer surface. Following grade 320 carborundum, then, a useful series of aloxite powders would be 225, 125, and perhaps 95, although the 125 will produce a surface fine enough to polish out in a few hours.

Another grading the amateur may encounter is the 'M' series used by the British American Optical Co. A common finishing emery from this supplier is M303½ (equivalent to aloxite grade 125), while M304 is about as fine as 95 aloxite. Finer powders are available, such as grade 50 aloxite, or M305 emery, but the use of ultra-fine powders does not seem to cut the polishing time required in proportion to their size, while the chance of contamination and scratching is increased.

A useful series of abrasives for grinding and smoothing a 15cm mirror would be:

80 carborundum	500g (1 lb.)
120 carborundum	100g (¼ lb.)
220 carborundum	100g (¼ lb.)

320 carborundum	150g (2 oz.)
225 aloxite	50g (2 oz.)
125 aloxite	50g (2 oz.)
95 aloxite (?)	25g (1 oz.)

Pitch

Pitch is a by-product from the distillation of wood or coal; it is a 'soft solid', which means that it flows steadily over a period of time, but is shattered by sudden force. The flowing nature of pitch is most important, for the curve or *figure* of a mirror changes during the polishing process, and the polisher needs to be able to follow suit.

Coal pitch is rarely encountered today because of the greatly diminished amount of coking taking place; hence, the amateur is almost certain to find himself using wood pitch. Swedish pitch, obtained from the pine tree, is commonly used. The hardness of the pitch is of great importance, and this point will be considered in due course. 500g (1 lb.) of pitch will be ample.

Rouge

The original ferrous oxide rouge, used by opticians right up to the present century, has now been largely replaced by mixtures of oxides of various elements, particularly the 'rare earth' elements such as cerium. These mixtures polish glass more quickly then the older type of rouge, and lack its wretched staining properties, so that change is a wholly welcome one. The type known as 'Cerirouge', grade E, is much used for fine optical work, but other proprietory types may be as good. Very little is needed, and 25g (1 oz.) of Cerirouge would polish two or three mirrors.

This short list concludes the items that will have to be purchased from specialist suppliers. Everything else that is needed will either be readily available, or very easily obtained from the high street. So let us start grinding!

CHAPTER FOUR

The Mirror: Grinding and Smoothing

The optical shop

The problem of where to work is the first question facing the would-be optician. Ideally he would choose a cellar, because subterranean conditions enjoy an almost constant temperature throughout day and night, and optical work is hindered by bouts of heat and cold. This is particularly so during the polishing and figuring process, because pitch polishers have their hardness changed noticeably by temperature fluctuations of as little as 2°C. Furthermore, cellars being draught-free, optical testing can be carried out spared from the flickering air-currents – to say nothing of floor vibration – that characterise an ordinary room, even with the door and window closed. Finally, cellars generally being neglected regions of the house, the inmates are not likely to protest unduly at having it taken over for an indefinite period!

These days, however, cellars are rarely to be found, and one must consider the alternatives. High on the list of priorities is being able to leave everything set up and ready for the next spell of work – hence, any spare room would be acceptable. If this is not possible, then the garage, or a garden shed (probably the worst possible choice because of dust, and temperature variations), may have to be selected as the only possible refuge. Dust and temperature changes are the prime enemies of the mirror-maker, and if the circumstances of his optical shop do not provide a natural deterrent, he must improvise. Polythene sheeting is cheap, and forms a perfect tent under which the work stand can be placed; and an eye on the thermometer will indicate when the temperature is near the selected standard for which the pitch was prepared

– somewhere between 15 and 20°C. is probably best.

The initial coarse grinding need not – and preferably should not – be done in the place chosen for smoothing and polishing operations. Carborundum powder is probably a more dangerous contaminant than ordinary household or even garage dirt, so do not foul your nest at the outset by bringing grit into it! The first two or three grades of carborundum work can be done very well in the open air; freedom from airborne dust does not become important until grade 320 carborundum is reached.

The grinding stand

A grinding stand, around which it is possible to walk, is needed, and it must be really solid and shake-free, since considerable pressure will be put on it. An old oil drum is ideal, since it is heavy to start with and can be made more so by pouring water into it. Alternatively, make a tripod stand of heavy timber with plenty of cross-bracing and a platform inside the legs just above ground-level on which heavy weights can be placed to stop the stand from rocking to and fro. A working height of about 85cm will suit most people. The top of the stand carries three small wooden blocks inside which the mirror or tool fits tightly enough so as not to move during the work, and it is helpful if the whole top is removable for washing-down between grades.

Storing the abrasives

Another job, best done before starting, is to bottle the abrasives. The grades from 220 carborundum down to the finest emery are most conveniently dispensed from plastic detergent squeeze-bottles. Put the abrasive in the bottles and add enough water to make the mixture sloshy; it will then be easy to shake it up and apply it to the work without fear of the contamination that attends open jars and spoons or brushes. A word of advice: when bottling, having got everything relevant clean, including your hands, start with the finest powder and graduate to the coarsest, so that there can be no carry-over of contamination. Grades 80 and 120 carborundum are too coarse to remain in suspension, and are best stored in screw-top glass jars with holes punched in the lid from the inside outwards; this prevents clogging in damp conditions. Each bottle and jar must be labelled, and stored so that they cannot possibly contaminate each other (coarse grades should be kept low down, so that stray grains cannot fall on to finer containers beneath). The relevant bottle is kept on a newspaper-covered table near the grinding stand, together with a squeeze-bottle of water in case the grinding mixture becomes too thick.

Bevels

Before commencing work, examine the mirror disc and decide which surface is to be ground. Plate glass is equally flat on both sides, but low-expansion cast discs usually have noticeably irregular surfaces, and some of these contours may be deep enough to represent an hour or two of otherwise unnecessary grinding. Examine the overall curve of the two sides and decide which one approximates most closely to a concave shape; other things being equal, this is the one to choose. Bubbles are another objectionable hazard, but it is difficult to tell how close they are to the surface unless the side of the disc is transparent. Having decided, take a medium-fine carborundum stone and grind a 2mm bevel all round the edge, at a 45° angle. The direction of the stroke should be from the face to the edge, but somewhat slanted so that the stone passes over a short arc of the disc's edge. Take the first stroke very lightly, as this is the one most likely to break chips of glass from the edge; such chips look ugly, besides acting as bastions for stray grains of coarse powders. Once the bevel is established, heavier strokes can be taken. This bevel must never be allowed to become ground out, because a sharp edge will tend to chip, leaving a flake of glass free to cause a disastrous scratch; hence, it must be renewed as soon as it becomes noticeably narrow. A bevel of the width advised should last through most of the work, unless the edge is from a very irregular casting, in which case the high spots will grind away rapidly.

The tool disc must also be bevelled, but it is not worth doing so until the grade 80 work has been finished because the margin is being ground down so rapidly. After this point, the maintenance of the tool bevel is as important as is that of the mirror.

Grinding strokes

Sprinkle some 80 carborundum on the tool's surface, add a few drops of water, and lower the mirror disc on to it. If the mirror's surface has a noticeably uneven face, take the first few minutes' grinding at an easy pace, without any very long strokes, until the high spots have been worn down. Otherwise – particularly if the high spots are at the edge – serious chipping could occur.

This apart, the method of rough grinding is to make the mirror disc move so that its centre bears as much as possible on the edge of the tool. When this happens, the pressure exerted by the operator on the back of the mirror is concentrated on to two regions, these being the centre of the mirror and the margin of the tool (Fig. 10); hence, the tool tends to become convex, while the mirror turns correspondingly concave. This effect can be achieved simply by moving the mirror across the tool, centre over centre, with a total stroke length equal to about 2/3 of the disc's diameter, the maximum deforming effect then occurring towards the end of each stroke; but it can be accelerated by allowing the mirror to move also somewhat sideways, so that some of the strokes are more along a chord than along a diameter of the tool. Under these conditions it can be seen that there is overhang of the mirror almost all the time instead of just at the extremities of the stroke. This is known as a 'W' stroke, because of the path traced out by the centre of the mirror disc (Fig. 10).

If this was all that the operator did, the convexing and concaving action would occur in only one direction, so that

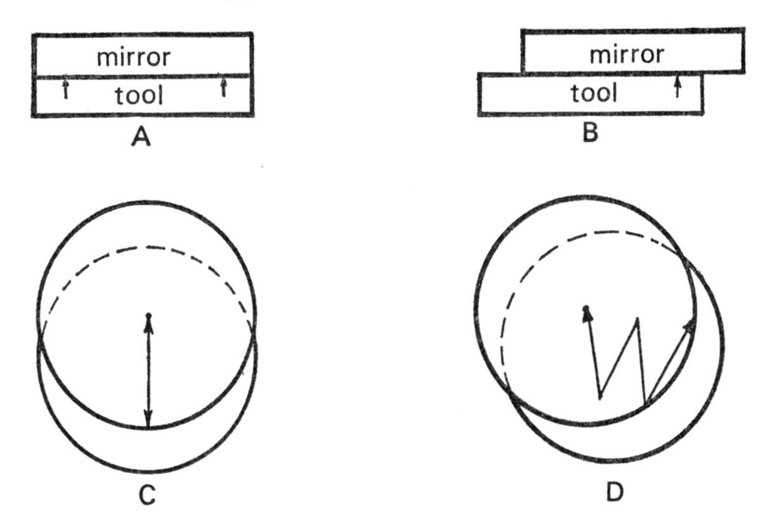

Fig. 10 Mirror grinding strokes. A: mirror and tool in the central position, with no localised wear. B: mirror overhanging tool, with curving tendency. C: centre-over-centre stroke for normal work. D: W-stroke, for exaggerated central wear on mirror and edge wear on tool

the mirror and tool achieved cylindrical curves. To achieve true 'surfaces of revolution' – curved surfaces that have the same shape when measured across all diameters – it is necessary to rotate the discs as well as to slide them. Tool rotation is achieved, in effect, by the operator walking slowly around the stand as he works, one revolution in 20 or 30 seconds being adequate, while at the same time the upper disc is given a slight rotatory twitch with the fingers during each stroke. There is no need to worry about precision of rotation; in fact, regularity is something to be avoided. The more random the motions, the more surely will a surface of revolution be achieved. Equally, there is no need to *worry* about being random – human nature will take care of this!

These three motions – back-and-forth stroke, revolution around the tool, and rotation of the upper disc – are the fundamental action of all grinding, polishing and figuring processes, and they will have become completely automatic well before the first 15cm mirror is completed. The only variation is in the nature of the stroke, which may be long or short, W or diametrical – the other two motions are always there.

It is interesting to point out, as an aside, that this virtually automatic method of curving a disc was not discovered until long after opticians were faced with the problem of producing curves on glass discs. In the old days they would take a disc of lead and beat it with a hammer until it was roughly of the curve required, then grinding the glass on the tool thus formed until it had taken the curve. Lead being very soft, periodic beating was required to renew the curve. Furthermore, fast-cutting carborundum was not manufactured until 1898; previously, the much slower-acting emery, or even sand, was used as an abrasive for roughing-out, and a long and tedious job it must have been!

Abrasive charge and 'wets'

The amount of carborundum and water to add will be determined by experience. Too much water produces a runny soup of abrasive and the discs will slither over each other without producing a crunchy grinding sound. Not enough water makes the powder collect in dry patches, and the discs do not slide properly at all. Too much carborundum powder inhibits the grinding process by cushioning the action of the grains. Ideally, there should be just enough to give a one-grain layer between the glass surfaces.

The rate at which the powder grinds down depends mainly on the pressure applied. As little as one minute's work may reduce the feel from one of coarse abrasion to that of barely-detectable cutting. When this point is reached, it means that the grains have been swamped by the ground-off glass. Both surfaces are then wiped off with a sponge, which is kept in a bucket of water, and a new charge of abrasive applied. It will probably be found that the surfaces are sufficiently wet from the sponging. Such a cycle is called a *wet*. It will be found that the duration of a wet becomes longer the finer the powder used.

Evenness of wear

After half a dozen wets, the operator's curiosity will demand an examination of the surface, to see what is happening. Ignoring local deformities should the disc be a cast one, it should be found that the centre has received much more grinding than the edge; ideally, the mirror's margin will be barely touched by abrasive pits, while the centre is heavily ground. The relative conditions of centre and margin tell us all we need to know about the stroke used. If the mirror is grinding evenly all over it means that the centre is not receiving enough preferential wear, and the concaving action needs to be accelerated, either by lengthening the stroke or widening the 'W'. If the margin is untouched, however, a central hollow is being produced but the wear is too localised, due to too extreme a stroke. This is a serious matter, and the writer would take issue with those authors who advise using extreme strokes in order to hollow out the centre of the disc rapidly with no thought as to whether the curve is a smooth one to the edge – leaving the smoothing process to sort out any irregularity. Although such a method may, technically, save time, it requires a good deal of experience to judge when an irregular curve has been properly smoothed out, and this experience is just what the beginner lacks. In any case, the curve on a 15cm f/8 or f/10 mirror is so quickly produced with reasonable strokes that there is no point in trying to take short cuts. It is far better for the worker to aim at a true curve, to the edge of his disc, from the very beginning. Very little time will be lost in the rough grinding, and much time – hours of it – may be saved later on, when the fine smoothing process finds that it is trying to smooth out irregularities of curve produced by the coarse powders. In the writer's experience, based on his own early efforts and examination of the work of amateurs, irregular coarse grinding and inadequate smoothing (the latter being partly dependent on the former) are the foundation of later troubles. A first-class mirror cannot be produced if the early work is not done perfectly, because mistakes made at this stage cannot be rectified later on.

The sagitta

The time will soon come to examine whether the curve is deep enough, and initial judgment can be made by measuring its depth directly, holding a straight-edge across a diameter and measuring the central depth or *sagitta* (S). The required value is obtained from the formula

$$S = \frac{r^2}{2R}$$

where r is the radius of the mirror, and R is the radius of curvature of the imaginary sphere of which the mirror's ground surface forms a part. The radius of curvature of a mirror being equal to twice its focal length, we have, for our 15cm f/10, values of 7.5cm for r and 300cm for R, giving

$$S = \frac{7.5^2}{2 \times 300} = 0.9\text{mm}$$

(For an f/8 mirror, S would be 1.2mm.) Therefore, a millimetre scale held against the straight-edge should show just less than one division between the bottom of the straight-edge and the centre of the mirror's surface. This rough test is enough to show when a more accurate check is desirable.

Measuring the radius of curvature

To do this, we establish the radius of curvature of the mirror by optical means, and the theory is shown in Fig. 11. If a light source is placed at the centre of curvature, the rays reaching the mirror will all be striking the surface perpendicularly or *normally*, and will therefore be reflected back along the same path, forming an image of the source on the source itself. If we move the source slightly to one side, the image will move in the opposite direction and can be caught on a screen. The mirror, then, is set up on edge, preferably in the testing stand described on page 80, and its surface is

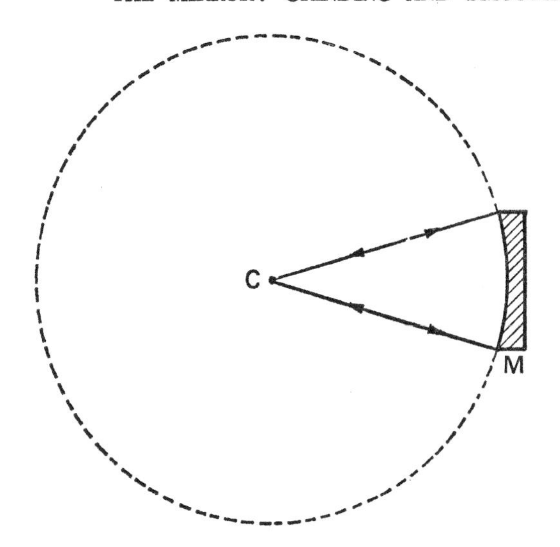

Fig. 11 Finding the centre of curvature of a concave mirror. The mirror M effectively forms part of the dotted circle whose centre is at C

splashed with water to make it reflective. A small piece of window glass has one side ground with 320 carborundum to form a screen, and the image of a torch bulb formed by the mirror is caught on this screen, the torch and screen being held at the same distance from the mirror and moved together back and forth until the sharpest possible image is obtained. The ripply water surface will give only an ill-defined image at best, but the focal position can be established to within ±5cm with no difficulty, particularly if an assistant is on hand to keep the mirror's surface wetted and told to hold one end of the measuring tape.

When the radius of curvature is indistinguishably close to 300cm, and the surface shows even grinding over the whole area (though the margin need only be slightly pitted), give a final two or three wets with a short stroke centre over centre to 'clean up' the curve. Then, subject to the checks mentioned below, the rough grinding is completed. The 80 carborundum jar is stored away where its contents can do no damage, and everything intimately associated with the rough grinding, particularly the stand, water bucket, and table top, is cleansed. Everything that the worker may have touched with his contaminated fingers is washed or replaced, as appropriate.

Cleaning up

The process of cleaning up between grades of abrasive really amounts to a declaration of banishment; on a large job it can be a major process. When George W. Ritchey (1864–1945), the eccentric genius who worked on the great mirrors of the Mount Wilson Observatory, came to the end of grinding the 100-inch mirror, his cleaning up preparatory to polishing included complete re-painting of the optical shop; it is recorded that the opticians making the 200-inch mirror for the Mount Palomar telescope took over 3 months to rid the workshop of all contamination! On the other hand, it must be realised that the home optician is unlikely to be able to control all extraneous sources of grit, and so, once he has washed, scrubbed, or thrown out everything contaminated, the policy henceforth must be that of containment. The working assumption to be adopted is that everything not known to be clean must be dirty – this includes the worker's hair and clothes! Furthermore, it should also be remembered that scratchy dirt is heavy, and therefore is potentially dangerous when somewhere above the optical surfaces, but harmless when on the floor. Dirt falling from ceilings can be intercepted by the polythene sheet, as already mentioned, and dirt down below can be ignored as long as no draught blows it into the air. The region immediately above the discs is the real danger area; therefore we do not lean over the work any more than can be helped, or hold an abrasive container (which may carry grit from the surface on which it was standing) so that its base can drop particles on to the glass. Carry things *round* the work rather than *over* it. The operator's hands are, of course, the most insidious agents of contamination, and a bucket of water should be handy for a quick rinse should they have had to touch anything. The habit should be formed of never wiping them on anything that may be present later during a finer grade of abrasive. A roll of kitchen paper is invaluable in the optical workshop.

These precautions may seem fastidious to a degree. The answer is that they should be! It needs only one grain of coarse carborundum or dirt to come between the fine-ground surfaces for a scratch to be produced that can involve back-tracking two or three grades (and as many hours' work) for its removal. Serious scratching is unlikely to affect the initial smoothing with 120 and 220 carborundum if elementary care is taken, but it is never too early to learn the habit of microscopic cleanliness. It soon becomes as automatic as the grinding action, and no further conscious thought will need to be given to it.

Checking for contact

The aim of smoothing is rather different from that of rough grinding. We have a mirror ground to within small limits of the required curve; the object now is to grind the surface as evenly as possible so that the coarse pits are everywhere fined down. Short strokes therefore replace the long strokes of the roughing process, and the 'W' augmentation is abandoned. By this means the margins of the mirror receive almost as much abrasion as the centre, and an examination of the mirror after a few minutes' work should show a finer overall grey in which the scattered deeper pits from the 80 work stand out in still relative profusion. If the margins of the mirror show no sign of action by the 120 grains, it may be because the stroke is too long, and further work with a shorter stroke should be undertaken; if the edge remains wholly coarse in texture, it means that the roughing-out was done too violently, and a curve has been produced that is not continuous to the edge, which is out of contact with the tool. It would be possible, eventually, to grind the curve out to the edge with the 120 grade, but a quicker way may well be to return to grade 80. The moral is to make very sure that the original grinding was carefully done. One way of checking this, during the first stage, is to make a pencil mark across a diameter of the mirror and to rub it, dry, over the tool with short strokes. In theory, if the mark wears away evenly, we know that contact is good over the whole surface; but the test is not conclusive, for even a good grinding technique produces the illusion of a slightly flattened edge since the two surfaces are not in physical contact, but are separated by an abrasive mixture of finite thickness. A better way is to grind a couple of wets of grade 120 before finally cleaning up from the 80 grade, to check that the two surfaces are in overall grinding contact.

Face-up working

Short stroke notwithstanding, face-down grinding is certain to wear the mirror's centre away faster than the edge, and a considerable saving of time can be achieved by reversing the discs for part of the time during each smoothing grade, thus concentrating the wear on the mirror's edge rather than on its centre. This not only accelerates the removal of marginal pits, but also has the most beneficial effect of ensuring a smooth curve right up to the edge of the disc through the positive action of the face-down tool. It should never be forgotten that the outer regions of a mirror are of far greater relative importance to its final performance than is the centre; half of its theoretical reflecting power is concentrated in the outer 29 per cent annulus, and the service rendered by the central regions is reduced still further by the 20–25 per cent obstruction of the Newtonian diagonal. In general terms, then, a little edge is worth a lot of centre, and the mirror's eventual quality will be dictated largely by the performance of its outer zones.

It will be worth repeating the splash test after perhaps 15 minutes of actual smoothing, since the finer surface will allow the radius of curvature to be measured to within ±2cm or so. Such accuracy will suffice for making up the telescope's tube and fittings. If the curve turns out to be still some way from the desired radius, grind with a somewhat longer stroke in order to flatten or deepen the mirror as required, but be sure that such work is followed by 15–20 minutes of short-stroke work in order to even up any irregularity of curvature caused by the long-stroke work. The 'W' stroke should certainly not be used at this stage.

Checking for pits

After half an hour's work, the mirror's surface should be examined to see how the pits are being removed. Assuming good contact all over, the vast majority of the surface will appear of even fineness, but examining the mirror using a magnifying glass or positive eyepiece of about 25mm focal length will reveal occasional deeper pits in the 'grey'. Some of these pits are caused by the occasional extra-large grain in the 120 abrasive – these are unavoidable, and will be removed in the next stage – but among these may be still deeper marks from the 80 grade. Indicate the position of the deepest pit visible by marking a small ring in waterproof ink on the back of the mirror, and continue grinding until it has vanished. We then know that all the remaining pits belong to the 120 grade. Good contact between the tool and mirror surfaces is proved by consistent quality of grinding from the centre to the margin of the disc. If the edge is noticeably more 'pitty' than the centre – a common defect after over-long face-down strokes – do some more work with the mirror face-up. *Never* delay treatment of a visible defect for the next grade. Remember that the general grey of the present surface, which appears so smooth and regular, itself hides deep pits that will appear as soon as the subsequent abrasive starts its work. What, then, will it make of any left-over marks from a grade two stages coarser? One is doubling or trebling the work for no more than an illusory advance. The importance of such discipline cannot be over-emphasised. In any case, a total of an hour's work with the 120 grade should be ample.

Fine smoothing

There is little need to go into detail over the rest of the smoothing process. By the time the 225 aloxite grade is finished, the surface will be fine enough to give a value for

R of $\pm$1cm or better on the splash test, and subsequent operations will barely change it at all.

It is during the last two grades of smoothing, and particularly the very last of all, that the danger of contamination rises to its peak. Ironically, it is even more serious here than during the polishing, since the pitch polisher contains channels into which a coarse particle may find its way, quite apart from the possibility of it sinking harmlessly into the pitch layer. During the 125 stage, the two glass surfaces approach each other to within a distance of only a few wavelengths of light, and there is nowhere for a grain of dirt to go, save over the edge – by which time it may have left a long scratch. Such a grain might do no harm at all during the coarser stages, when the surfaces are more widely separated. It is clear that the greatest peril lies at the moment of placing one disc on the other at the beginning of the wet. Therefore, the freshly-applied abrasive should be spread out over the lower surface with a clean finger, feeling for any particle, and the other disc then lowered centrally on top and moved very gently from side to side, listening and feeling for any crunchy sound; if anything is heard the disc should be raised, with no sliding motion, and the charge cleaned off. Contamination is unlikely to occur once the wet has started.

Seized-up discs

The wets of these fine powders will last for as long as 5 minutes, so that 8 or 10 wets of each grade may be sufficient; but inspection is the only trustworthy guide. Another danger is of the mixture drying up during the wet when the powder has become worn very fine. When this happens, the glass surfaces may seize together. If the mixture is thin and drying, squirt a few drops of water on to the work before sliding the discs apart. A long final 125 wet, keeping the discs going with water but adding no fresh charge of abrasive, can leave a very fine surface for the polisher; but the danger of seizing as the abrasive wears down must be borne constantly in mind.

If the discs do stick, do not attempt to part them by force, since scratching or flaking may well occur. Instead, submerge them in a bucket of clean water and allow osmotic pressure to draw the water in between them, when they will fall apart. If this treatment fails on its own, a gentle tap on one disc via a piece of wood may transmit sufficient shock to break the contact. Heating of the water has been recommended as an antidote for firmly-stuck discs, and, in extreme cases, the desperate optician will adopt his own methods, praying that his hard-won optical surface emerges unscathed.

The final check

A well-smoothed glass surface is a fine sight. Assuming that the back of the disc is clear, the glass will appear translucent and much detail can be seen through it. The filament of a clear-glass lamp, only dimly and redly visible at glancing incidence on a grade 80 surface, can be seen reflected at a 45° incidence, and a brief rub of the surface with a finger will polish away enough of the grey from a small spot to allow normal reflection to occur. On the other hand, it should not be supposed that these properties prove conclusively that the surface is in a fit state for polishing. Only the magnifier can do that. A 25mm lens should show the grade 125 surface as the finest stipple, with no pits showing any depth, and no degradation of quality right up to the mirror's edge. It is now that any untreated coarse pits from the early grades will show up in all their hideous prominence. Their absence reveals a job well done, and the optician can have confidence that he could not have a better springboard from which to launch into the polishing and figuring processes.

CHAPTER FIVE

The Mirror: Polishing and Testing

The polishing process

The grinding process is one involving shattering (carborundum) or shaving (aluminium oxide) of the glass surface. Both processes can occur only when the surface of the tool is hard and unyielding. Polishing is a different matter altogether, as it depends on the action of a yielding surface, and pitch is the material almost exclusively used for the purpose. The exact process by which a glass surface becomes polished is obscure. Is it a microscopic form of fine grinding, in which individual molecules of glass are worn away, or is some sort of glass flow involved as well, the outer skin becoming effectively molten through friction? There is certainly no doubt at all that glass is removed, for a disc becomes thinner as polishing continues; but there are objections to the 'fine grinding' idea, for rouge will not polish glass if fed between two fine-ground glass surfaces, while fine aloxite will produce a passable polish if used on a pitch polisher. Therefore, the polisher itself plays a fundamental role in the business.

It is probably true to say that making a polisher has given amateurs more headaches than any other part of the mirror-making process, figuring excepted. Part of the reason is the old problem of book-learning against direct instruction. Another factor is the reluctance of people to experiment with new methods, outdated advice being passed on without further thought.

The hardness of the pitch

The object of the exercise is to coat the tool with a uniform layer of pitch, which therefore corresponds to the curve of the mirror, and to cut this layer into squares or facets to allow the pitch to flow freely under the pressure of polishing. The first action concerns the hardness of the pitch, which can vary, as it comes from the supplier, from the consistency of toffee to that of coal. Take an old saucepan, melt the pitch in it (preferably out of doors, to stop the smother pervading the house), stir it up well, and cool a small sample in water at the likely temperature of the optical workshop. This is most important, for the hardness of pitch changes noticeably over a temperature range of a very few degrees; 'summer pitch' will almost certainly be too hard for winter working, and conversely 'winter pitch' will flow too rapidly on a hot summer's day. It is for this reason that a cellar simplifies the optician's problems.

To judge the hardness of the pitch, take a penny, rest the elbow on a table, and allow the weight of the arm (typically about 500g) to press the rim of the coin into the sample, noting the time taken to leave a dent 6mm long. We can thus classify the pitch, in terms of hardness, as '15-second', '30-second', etc. At this point we enter the realm of personal opinion as to what is a satisfactory hardness for a polisher, but few people would dispute that 15-second pitch would classify as an ultra-soft type, of little use except for making up sub-diameter polishers for figuring (see page 97); in normal working it will flow so fast that much time is spent re-cutting facets. The writer would advocate 45-second pitch as being most suitable for a normal polisher. Very hard pitch is difficult to work into shape, tends to produce sleeks more easily, particularly on plate glass, and actually polishes more slowly than the medium or soft variety.

If the pitch is too soft, which will almost certainly be the case if it is untreated Swedish pitch, the most common type available in this country, it must be hardened by evaporating off the lighter constituents. Two or three hours' steady heating will be required to make much difference, but do not forget to allow the sample to cool thoroughly before testing. If the pitch is already too hard, add a small amount of pure turpentine (not white spirit), and stir it into the melted mass very thoroughly before making another test. Not more than a teaspoonful at a time should be added to 500g of pitch, for a lot of heating will be required should the mark be over-shot!

Making a polisher

The making of the polisher resolves itself into the problem of establishing the pitch, the tool and the mirror at their correct respective temperatures, and it is here that experience helps. A method often advocated is to make a mould by running a ring of paper around the edge of the tool, pouring molten pitch into it, and allowing it to cool sufficiently to stand up when the ring is removed. The warmed mirror is then worked over the flat pitch surface until it has acquired the correct convex curve. What happens in practice is that the air bubble between the mirror and the centre of the polisher prevents rapid contact from being achieved, so that the pitch and the mirror – particularly the former – may well have already cooled beyond the point where any mobility remains, necessitating re-heating and a lot of extra work. The trouble clearly lies in starting pressing operations on a flat pitch surface. If the pitch can be poured to correspond approximately to the convex shape already possessed by the tool, a minimal amount of pressing will be required.

A better method to adopt is as follows. Place the tool in a bowl of hand-hot water – pitch adheres better to warm glass than to cold – and stand the mirror in front of an electric fire. When ready, it should not be too hot to be held comfortably in the hands – if of plate glass, a somewhat lower temperature is advisable because of the possibility of breakage should a drop of cold water fall on it. Meanwhile, the pitch has been well melted and is now allowed to cool, while being stirred constantly with a wooden stick to keep the colder layers near the side of the saucepan mixed in with the rest of the mass. Periodically lift the stick and observe the way the stream of pitch runs back into the melt. If it sinks instantly into the molten mass, it is still too hot; when it retains its identity for a couple of seconds before melting away, the temperature is about right. At this point, the tool and mirror must be at their correct temperatures as well.

Now dry off the tool, lay it on a level surface (this is really a kitchen-table job), and pour a steady stream of pitch on to its surface with a spiral motion, beginning at the centre and working steadily outwards. If all is well, the pitch will maintain a layer about 3mm thick. Do not worry if parts of the spiral do not meet up in a continuous layer; such gaps do not matter, and they may in any case fill themselves in during the pressing. Run the pitch right out to the edge of the tool; it does not matter if some flows over the edge. Now place a dry finger on the marginal (softest) pitch; if it sticks to the skin, it is still too hot for safe pressing. When it has cooled enough to take an easy imprint without being tacky, smear the hot mirror generously with rouge and water mixture – best kept in a flip-top plastic carton and applied with a brush – and lower it over the polisher, moving it continuously so that the pitch has no chance of adhering to the hot glass. Observed through the back of the mirror, the places of initial contact will be seen as a spreading blackness. Keep pressing and moving the mirror until the blackness has spread as far as it will go. Then slide the mirror off.

If everything was just right, the pitch will have pressed into an even, flawless convex surface. If there are large areas still out of contact, further warm-pressing is required, but small depressed holes or bubbles are of no account as the pitch will flow into them during subsequent pressing or polishing. If no major re-pressing is indicated, allow the mirror to cool; otherwise, heat it up again, and cut grooves in the high (contacted) regions to help them sink down to the level of the rest when the warm mirror is replaced. When the surface is satisfactory, it is time to cut facets in the pitch.

The facets

The facets perform several functions. They leave room for the rouge-water mixture to circulate, allowing polishing to continue for long periods. They help the polisher to stay in contact with the mirror by letting the pitch flow freely under the warmth and pressure of polishing without creating any local build-up of material. Finally, they accelerate the polishing process through the fact that polishing takes place most vigorously at the edges of the facets. Therefore, the more facets the better. Their shape is immaterial, provided they are distributed evenly over the face of the polisher, but square ones are the easiest to cut. The major facets on a 15cm polisher should be about 2cm square; they can be sub-divided later on.

The secret of successful facet-cutting is to have a very sharp blade, and to take small shavings rather than deep cuts. So invest in two or three *new* safe-edge razor blades, and keep pitch and blade wet to reduce the tendency for the cutting edge to become clogged and blunted with pitch. First of all, trim off the surplus material from the edge, shaving it away with a downward stroke to reduce the likelihood of chipping the working surface. A slight chamfer can profitably be applied, making the diameter of the polisher a millimetre or two less than that of the mirror, to facilitate the subsequent cold-pressing. Now shave out the major facets. There is no need to draw straight lines, or to measure them accurately; eye-estimation is adequate, and a little random variation will do no harm at all! Cut the grooves in the form of a 'V', first cutting out one side and then turning the mirror round and cutting the other side, going right down to the glass surface. It may take three or four separate cuts to shave down through the pitch without taking flakes out of the facets, but there is no need for hurry. The writer suspects that much of the flaking trouble encountered during this process is due to rapid and deep cutting in an attempt to take out the groove in one operation. However, it should also be added that a few flakes taken out of the surface do no harm provided they are not localised at one particular radius of the polisher, when the reduced polishing action could possibly produce a high zone on the mirror.

At this juncture we encounter an old dogma that should have been discredited long ago. It has been repeated almost *ad nauseam* that the network of facets on a polisher should be placed eccentrically with respect to the polisher's centre, as otherwise the mirror will polish in zones or rings. The only way in which this could happen would be if the polishing strokes, and the outlines of the facets, were so precisely ordered that the gaps between the facets returned to exactly the same zones of the mirror during the strokes. With typical groove widths of 2–3mm, and the irregularities inevitable in manual cutting of facets and hand-polishing, this could not possible happen. The cause of zonal irregularities is bad contact of the polisher, a point to be returned to later.

During the few minutes taken to cut out the main facets, the warm pitch will have slumped somewhat in the centre because of the extra warmth maintained there. In other words, the radius of curvature of the polisher will have lengthened, and contact must be restored before sub-faceting and polishing; everything must also be allowed to cool down to the working temperature. So the still-warm mirror is smeared with rouge, and the polisher, brushed free of pitch shavings, is set down on top of it. We do it this way so that the pitch tends to flow *into* contact at the centre. Run a band of electricians' PVC tape, or other waterproof material, around the edge of the discs to prevent them from drying out, and place a weight of a couple of kilograms on the back of the polisher. The sandwich can then be left overnight to cool down completely.

The Foucault test

With polishing imminent, it is time to consider the testing of the optical surface – for testing can begin as soon as the glass is reflective all over, which will take only a few minutes' work. The object of polishing is to produce as nearly perfect a spherical surface as possible, true to the very edge of the disc, on which parabolishing or figuring' operations can commence. We therefore, initially, require a test that will tell us if our mirror is spherical, and to within what limits.

We have to thank the French physicist Léon Foucault (1819–68) for devising a laboratory test that has stood both amateur and professional opticians in good stead for the past century. Before Foucault's time, telescope optics had to be tested on the stars, or at least on distant terrestrial objects – William Herschel corrected many of his mirrors by observing the towers of Windsor Castle, which were visible from his garden! George With of Hereford, another fine mirror-maker, tested his mirrors on an 'artificial star' – the reflection of the sun in a watch-glass – set up some hundreds of yards away. Such tests were cumbersome to arrange and depended rather critically on the weather; it testifies to their users' perseverance and skill that fine results were obtained.

Foucault's test requires only simple equipment, no more space than is needed to accommodate the radius of curvature of the mirror, and is sensitive to errors of figure of between 1/20th and 1/100th of a wavelength of light, depending on the circumstances. In other words, it is more sensitive than necessary. The reader should not, however, be surprised to learn that the test depends to a considerable extent on the judgment and experience of the optician, so that its effective sensitivity is to a large extent governed by the user's skill – the results are not laid out in black and white, but must be interpreted through what is seen, rather as a good astronomical observer must judge what is real and what is illusory in the flickering image of a celestial object seen through a troubled atmosphere.

Testing a spherical mirror

On page 60 we attempted to determine the radius of curvature of our rough-ground mirror by setting up a light source at one side of the centre of curvature and receiving its image nearby. Let us now suppose the mirror to be polished, and substitute a brightly-lit pinhole for the torch. If we place our eye at the image of the pinhole, we see light being re-

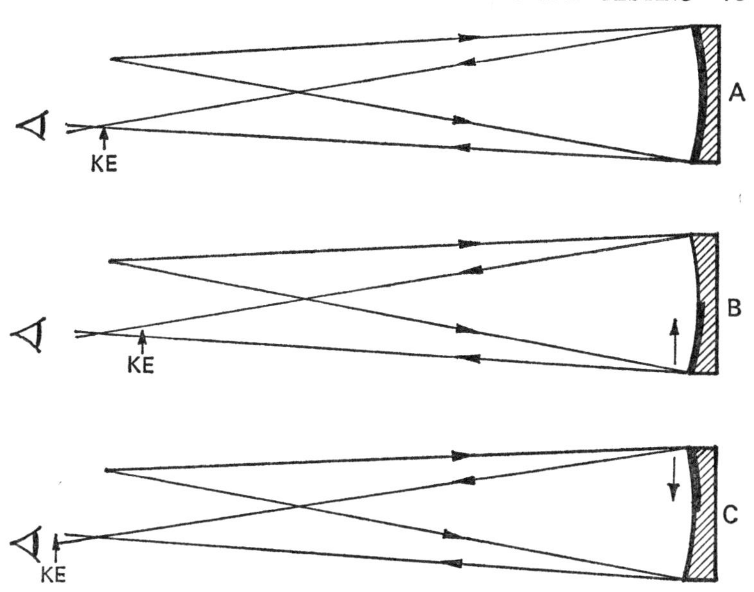

Fig. 12 Knife-edge test of a spherical mirror. The dark band on the mirror's surface indicates the position of the shadow as the knife-edge KE is passed across

flected back from all regions of the surface, and it appears uniformly illuminated.

If now the eye is withdrawn slightly, but so that the full illumination is still seen*, and a sharp edge (known by opticians as a *knife-edge,* regardless of its true nature) is passed across the image of the pinhole, the light is suddenly cut off and the mirror appears to go dark evenly all over. If the pinhole is small enough, this darkening is virtually instantaneous and uniform over the whole surface (Fig. 12a). But suppose instead that the knife-edge is moved towards the mirror slightly, so that it is inside the image of the pinhole (Fig. 12b). It now cuts first across the rays coming from the same side of the mirror as the knife-edge itself, so that the

*If the eye is withdrawn too far, the cone of light becomes larger than the observer's iris, so that light from the margins of the mirror is cut off.

observer sees a straight-edged shadow pass across the mirror in the *same* direction as the knife-edge is moving. Conversely (Fig. 12c), if the knife-edge is moved back to a position outside the image, it first cuts into the rays coming from the opposite side of the mirror, and the shadow appears to move in the *opposite* direction to that of the knife-edge. In this principle lies the germ of the Foucault test: by observing the direction of the shadow, we know where the knife-edge lies in relation to the centre of curvature.

Aspheric mirror curves

Let us now suppose that the mirror is not truly spherical, but has some aspheric shape. This means that the radii of curvature of the centre and edge differ. If the central radius is longer than that of the margin, the curve is said to be *oblate*, and represents the flat section of an ellipse (Fig. 13). Oblate curves are not often used in optical designs, and of course are quite opposite to the one wanted for our mirror. Passing through the sphere, we encounter the *prolate* family, with a short central radius. There is first the family of prolate

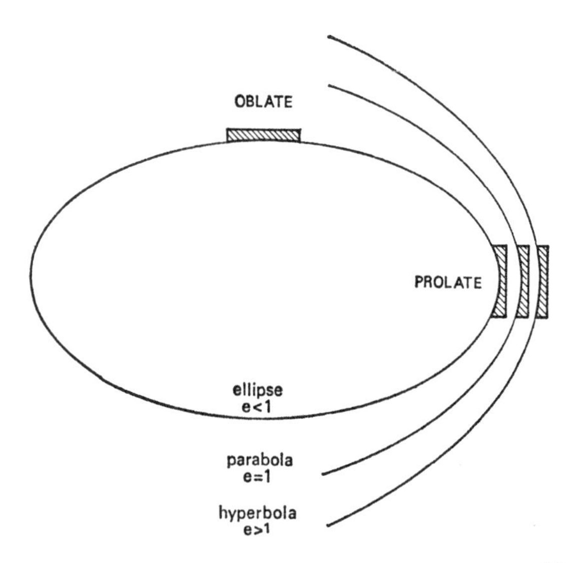

Fig. 13 Aspheric curves

ellipsoids, which correspond to the steep section of an ellipse, as shown in the diagram. The degree to which the central and marginal radii differ is a measure of the eccentricity (*e*) of the ellipse. A figure with $e=O$ is a circle (zero eccentricity); as *e* approaches unity the ellipse becomes more and more extended until we can imagine it as being infinitely long, so that the arms never meet up. Such a figure, with $e=1$, is the parabola, the shape required for a Newtonian mirror. When *e* becomes greater than unity the curve is said to be *hyperbolic*, the effect of which is to make the difference between central and marginal radii even greater than that of a parabola. The difference in radius of curvature between the centre and edge of an aspheric mirror is given by

$$R = \frac{e^2 \, r^2}{2R}$$

where r is the radius of the mirror, and R is the mean radius of curvature. For a parabolic mirror, where $e=1$, this reduces to $r^2/2R$, the same as the sagitta we measured earlier.

Testing an aspheric mirror

If the mirror is aspheric, it is at once clear that there can be no unique image of the pinhole at the centre of curvature, since each zone of the mirror has a different radius of curvature. In Fig. 14 a prolate mirror is being tested with Foucault's apparatus. At position *a*, the knife-edge is inside all the zones and a more or less straight crossing shadow is seen, moving in the same direction. It is then moved back to position *b*, the centre of curvature of the central part of the mirror. The shadow now darkens evenly across the centre, but still moves the same way as the knife-edge across the margin, giving the appearance shown. At position *c*, which represents the mean radius of curvature of the whole mirror, the knife-edge is outside the centre of curvature of the mirror's centre, but still inside that of its edge; hence two shadows are seen, moving in opposite directions as the knife-edge passes across.

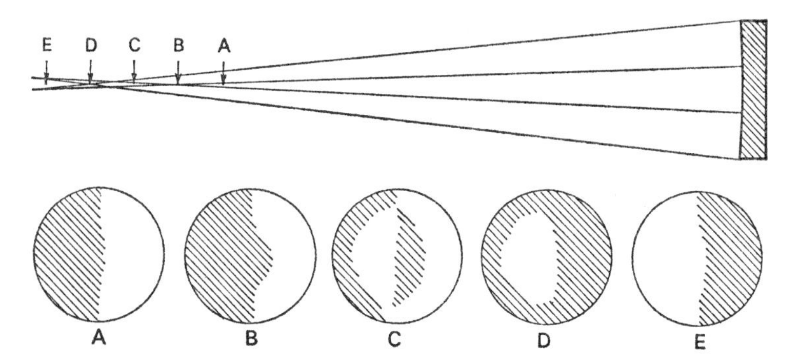

Fig. 14 Knife-edge test of a prolate surface. The shadings indicate the appearance of the mirror when the knife-edge has passed halfway across the beam of light from the mirror. The light source is not shown

At position *d*, the centre of curvature of the margin, the edge darkens evenly, while at position *e* a shadow is seen moving across the whole mirror in the opposite direction to the knife-edge travel.

If both pinhole and knife-edge moved together in this back-and-forth scanning, the distance A for the mirror would correspond to the measured travel between positions *b* and *d* if the mirror were parabolic. Normally, however, the pinhole is left at a fixed distance from the mirror and the knife-edge alone is moved through the different centres of curvature. This doubles the travel, and so halves the errors of measurement. Therefore, with a fixed pinhole, we have $R = e^2r^2/R$ *for* an aspheric mirror in the Foucault test, or r^2/R for a parabolic one. R is usually termed the *aberration* of the mirror, although this is a confusing name in view of the many other aberrations that concern opticians!

This wonderful test, then, allows the optician to *see* the optical shape of his mirror in terms of light and shade. It is extraordinarily sensitive. On page 26 it was pointed out that the surface difference between spherical and parabolic 15cm f/10 mirrors is 0.00015mm (0.3 of a wavelength of yellow light). With $r = 7.5$cm and $R = 300$cm we obtain a value for R of 1.9mm, corresponding to a magnification of over 10,000 times. (For a 15cm f/8 mirror, R=2.34mm.)

Making a Foucault tester

Plate 3 shows a typical Foucault set-up. The knife-edge can be a razor blade clipped to a flat-based stand. A reference edge is cut into the base, beneath the knife-edge, so that the various zonal readings can be recorded. The pinhole itself is made in a small piece of kitchen foil using a sharp needle. One good way of producing a tiny hole (a diameter of not more than 0.05mm is recommended) is to make a stack of three or four pieces of foil and lay them on a hard, smooth surface such as a piece of glass. Then take a *new* needle and, with a twirling motion, let it sink into the stack. With luck, it will leave a very fine hole in the last sheet it pierces. Examine the hole against the light with a lens or eyepiece of about 20mm focus; it should appear as a tiny disc, almost a point of light. Too large a hole will dilute the Foucault shadows and make it more difficult to define the various zones on the mirror; on the other hand, a microscopic hole will pass too little light to give an acceptable view.

Much depends on how well the pinhole is illuminated. A torch bulb will give a bright image only if the filament is exactly in line with the pinhole and mirror, and it will probably not give full illumination over the whole surface of the mirror. A diffusing screen of ground glass between the bulb and the pinhole will remove the need for such accurate positioning, but the brightness of the pinhole may be insufficient unless a very brilliant and hot lamp is used – a car headlamp bulb, for instance, which requires the nuisance of a special 12-volt supply. A method that the writer has found to give good results is to use a condenser system. Take two 1.5-volt lens-end bulbs, break the lens off one using a pair of long-nosed pliers, and mount it up with the pinhole near its focus, as shown in Fig. 15; the flatter of the two curves should be towards the pinhole. This is fixed at the end of a short tube. The other bulb is fitted, in its holder, at the other end of the tube. The light from the filament is then turned into

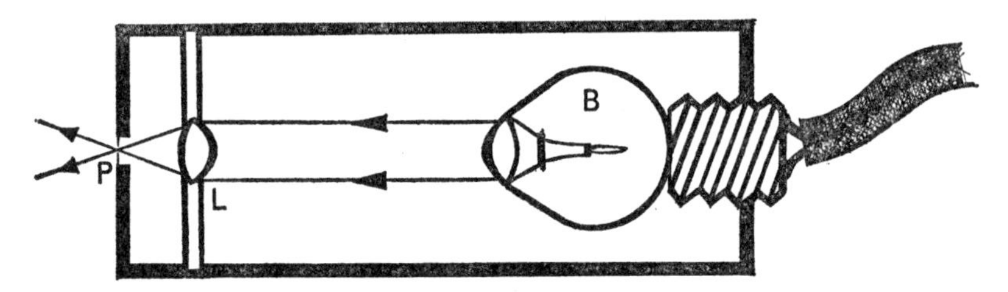

Fig. 15 An illuminated pinhole or 'artificial star'. B: lens-end bulb. L: condenser lens. P: pinhole

parallel light by the first lens and re-focused on the pinhole by the other (condensing) lens. Some adjustment may be required to bring the focused image on the pinhole, but eventually a brilliant artificial star will be obtained. The housing need be no more than 15mm across, so that the source and image can be located very close to the optical axis of the mirror. This is important, for wide separation will give the shadows an astigmatic or twisted appearance. A mirror of f/10 is, admittedly, insensitive to moderate departures from axial alignment, but compact sources will be appreciated if f/4 or f/6 mirrors are attempted later.

The pinhole is mounted on a stand so that it is as high as the middle of the knife-edge. All that is needed to complete the equipment is a simple wooden stand on which the mirror can be placed for testing (Fig. 16). The upright part carries three projections against which the back of the mirror is located, and its lower rim rests between two wooden blocks. If these are sloped, as shown, the weight of the mirror pushes it back against the location points and no retaining catch at the top of the disc is required. An adjustment screw at the front of the base is very useful for initial alignment. If the mirror is always returned to the stand with the same orientation, the rear projections will locate against the same points on the back of the disc and the image will return automatically to the same region of the knife-edge, saving much time and loss of temper in hunting for the elusive point of light! A torch will aid the setting-up of the test.

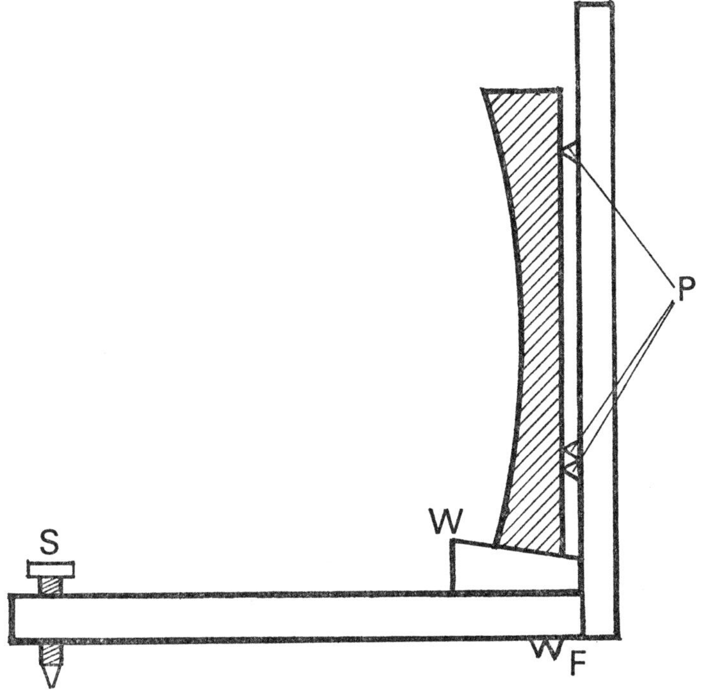

Fig. 16 A mirror testing stand. F: rear feet. P: rear location points. S: vertical adjustment screw. W: wedge-shaped retaining blocks

The importance of perfect contact

We can now return to the mirror and the business of polishing. If the sandwich has stuck, it can be separated by allowing water to percolate between the two surfaces, either using running water from the tap or by dowsing the discs in a bucket of water. It may be found that the facets have sunk so far that the channels are blocked, in which case they are renewed. Light sub-facets are then cut with two or three grooves across each major facet. The waste pitch is brushed off under a stream of running water, or into a bucket of clean water, and a brushful of rouge is applied. Polishing is then commenced, using short strokes, after allowing ten minutes of weighted cold-pressing to elapse following the cutting work. Every time the mirror is taken from the polisher, be it for re-cutting of the pitch or for testing, a ten-minute press should be given to ensure that perfect contact has been restored. At the end of a session, if it is known that further polishing is necessary, mirror and polisher are stored away as previously described, although a weight of only a kilogram or so will probably be enough to regain the contact.

This matter of contact cannot be afforded too much attention, for it decides how satisfactory the preliminary figure will be. The technique of grinding two glass surfaces together produces two matching spherical surfaces with an accuracy of considerably better than the average grain size of the abrasive; a well-smoothed 125 surface will have no departure from sphericity greater than a couple of wavelengths of light, while considerably more glass than this must be removed in order to reach the bottom of the deepest remaining pits*. Therefore, the polisher itself must assume responsibility for maintaining the surface in an effectively spherical condition, and it can do this only if its surface accurately matches that of the mirror. Gross lack of contact can be seen by viewing the facets through the back of the mirror, when any not in close proximity to the glass appear hazy; but *optical* contact may still be lacking even when this check appears favourable. The most tell-tale sign is the way the mirror submits to the the polishing action. If it sticks and binds as it passes across the polisher, we know that contact is confined to the edge, and there is a resisting grip; if it tends to swing around its centre, then edge contact is deficient. A mirror in perfect contact slides without any preferential motion.

Starting the polishing

An initial 15-minute polishing spell will tell much about the state of the polisher, particularly if a testing apparatus has already been made up, as it should have been. A 1/3 diameter stroke, with no side, is best, for the aim is the same as in fine grinding – to wear the whole surface of the mirror as evenly as possible. Try to grip the mirror around the back rim

*Even a well-smoothed surface is likely to contain occasional pits about 5 microns deeper than the rest of the abrasion.

rather than down the sides, since local expansion of the glass from the warmth of the fingers may, according to some authorities, produce excess marginal polishing, and hence a turned-down edge when the glass has cooled. The writer cannot claim to have ever noticed this effect, which is merely alluded to as a point for consideration; in any case it is not of critical importance, since subsequent face-up work, as recommended later, will remove any slight effect imparted at this stage. The harder the pressure, the faster the glass polishes – the limit being the softening effect of friction upon the polisher. After 15 minutes, the mirror should be fully reflective all over, though almost certainly better polished in the middle. If the margins are still dull, full contact has not yet been obtained and further cold-pressing is required. Do not go on polishing in the hope that contact will come: by the time it does, an optical pit may have been dug into the centre of the disc. A little care saves a lot of repair – as in all stages of mirror-making.

If the Foucault test can be set up, so much the better. Place the apparatus on a couple of solid tables, and locate the image next to the pinhole where it can be 'knifed'. We need not expect a perfect sphere at this stage, but there should be no aberration greater than a couple of millimetres. Of more interest than the overall figure is the presence of narrow rings of shadow, which indicate defective zones (see Plate 4). If these are present, something is seriously wrong with the contact, and more substantial pressing is required. If the surface appears even, all is well, and polishing can continue after normal cold-pressing. After a continuous spell of half an hour, the work should be stopped to allow the polisher and mirror to cool; if the sub-facets have started to fill up, they are renewed. Meanwhile, a further test of the mirror can be made. Periodically renew the chamfer on the polisher, to make its diameter very slightly less than that of the mirror; this will prevent a thin raised rim from being formed during the pressing.

Polishing rates

Polishing times vary so much according to the rapidity of the strokes, the pressure applied, and even the nature of the pitch and the rate at which the rouge is renewed (a fresh dab is added whenever the mixture starts to look pale and thin, but if it is simply drying out, a squirt of water will suffice), that only the most general indication can be given. The centre of the mirror will probably look completely clear after an hour, but an examination with a magnifier may well reveal a scattering of deeper 125 pits, appearing as no more than black specks – any pits showing real depth are left-overs from the coarser grades, and will not polish out in any reasonable time. Another hour could be taken up in ridding the centre of these specks. All this time, the edge will be polishing much more slowly. This is partly the deepening effect of the stroke's overhang, but there is also the deforming effect of the stroke on the polisher: the marginal pitch sinks slightly under the excess pressure, and so the radius of curvature of the polisher shortens. It is therefore trying to polish a curve of slightly shorter radius on to the mirror. Frequent pauses for cold-pressing will tend to alleviate this effect.

Face-up polishing

Once the central 8cm or so has polished out, it is, in the writer's opinion, an excellent idea to reverse the discs and polish with the mirror underneath. It saves considerable time, because extra wear is being directed at the region most in need of it, but there is another reason too. We have already seen that the quality of the outer annulus of the mirror is of paramount importance. One-half of its reflectivity is contributed by the outer 29 per cent ring, and a quarter by the last 13 per cent. A defective margin is therefore a serious matter, and the most common defect of all is a turned-down edge, in which the outer few millimetres have a curve of longer radius than that of the rest of the mirror. Face-down working, particularly if the stroke is somewhat on the long side, always tends to leave the edge behind in the radius-shortening process and hence, since we are concerned with the general radius of curvature of the mirror, effectively *lengthens* the radius of curvature of the edge. This relativity of wear of different regions of the mirror must be borne constantly in mind; only in this way can the worker understand how an untouched zone can change its shape! If we turned the mirror face-up, the positive polishing action at the margin now sweeps away the crest of glass and leaves a new and uniform curve true to the edge. Long strokes are not necessary; the reversal alone achieves the change.

It must be admitted that face-up polishing has earned little favour in the textbooks, the principal complaint being that sleeks and scratches are more likely to occur. Sleeks are spider's-thread scores, caused usually by the finest floating dust or impurities in the rouge, and in the writer's view will affect face-up and face-down mirrors equally, if they are present at all. Scratches will occur during polishing only through unforgivable carelessness or real bad luck; the great preponderance occur during fine smoothing. A good optician will, naturally, do all he can to avoid ugly blemishes appearing on the surface of his mirror; but he will be even more concerned with producing the finest possible performance. The argument that telescopes are made to look through, not at, may often be a last-ditch defence against a charge of poor workmanship, but nevertheless correctly emphasises the relative values of the case. The presence of a few sleeks makes absolutely no detectable difference to the performance of a mirror; a turned-down edge can ruin it. *Verb sap!*

Checking the figure

As the polishing nears completion, a more critical eye must be kept on the form of the optical surface; there is no point in cheerfully polishing away until the last pit has gone if some remedial action can be taken to bring the surface nearer to the desired final form without sacrificing polishing efficiency. We start, then, to measure any asphericity that may be present, and to do this a sheet of plain paper is taped to the testing table so that the knife-edge can be moved about on it. When the knife-edge has been located at the centre of curvature of a zone, a pencil mark is made against the locating edge beneath; subsequent zones also have their positions marked, and the distance between the lines can afterwards be measured, leaving a permanent record of the mirror's optical state at that particular stage in the work. The writer much prefers this method to running the knife-edge along a scale and taking measurements after each zonal examination; in the latter case there is no permanent impartial record, the relaxed eye must be forced to read minute divisions by torchlight, and, if a series of zones is being measured, a knowledge of the previous intervals can bias the later settings. An undeniable drawback of the Foucault test is its amenability to bias. Unless the mirror has a short focal ratio, when the test is unsuitable anyway, the shadows are nebulous, fading into each other across a hazy no-man's-land representing a knife-edge movement of a few tenths of a millimetre; the hand of bias can make the interval longer, or shorter, in response to the subsconscious desire to compensate for a known or suspected fault. But, working in the semi-darkness, the pencil marks are drawn blind and so are subject only to the human errors of judgment, which can be partly alleviated by repeating the measures two or three times and comparing the marks. If the room is bright enough for the marks to be seen, then arrange to test the zones in the order that brings the knife-edge base back over the marks, so hiding them – in other words, from the centre outwards for a prolate mirror. In this way one can have confidence that it is the stark truth about the mirror's surface, and not wishful thinking, that forms the basis of the analysis and strategy required in the figuring process.

CHAPTER SIX

The Mirror: Figuring

Figuring is a process in which there are two stages to be considered: diagnosis and cure. Clearly, no action can be taken until we are confident that we know the true shape of the optical surface, so the problem of diagnosis must receive first attention.

Choosing the curve

The writer well remembers experiencing considerable difficulty over this when making his first mirrors, due chiefly to his failure to realise that the apparent shape of the mirror's surface, as revealed by the knife-edge shadows, changes as the knife-edge moves forwards and backwards to different centres of curvature. In all cases we are observing high and low regions of the figure *in terms of the sphere corresponding to the radius of curvature of the knife-edge setting*, and therefore, although different zones of an aspheric mirror have the same difference between their radii of curvature no matter how we test them, we can make one or both appear convex or concave according to the position of the knife-edge.

Fig. 17 indicates the situation. Suppose that we are testing a prolate mirror. If the knife-edge is first set up so that the margin darkens evenly, it must be at the centre of curvature of this zone, so that dotted line *a* is our reference sphere. The central region, having a shorter radius of curvature, is therefore depressed below this line and appears concave, the shadow moving in the opposite direction to the knife-edge because we are outside its centre of curvature. To obtain a second interpretation, advance the knife-edge to the centre of curvature of the central zone. The reference sphere *b* now has

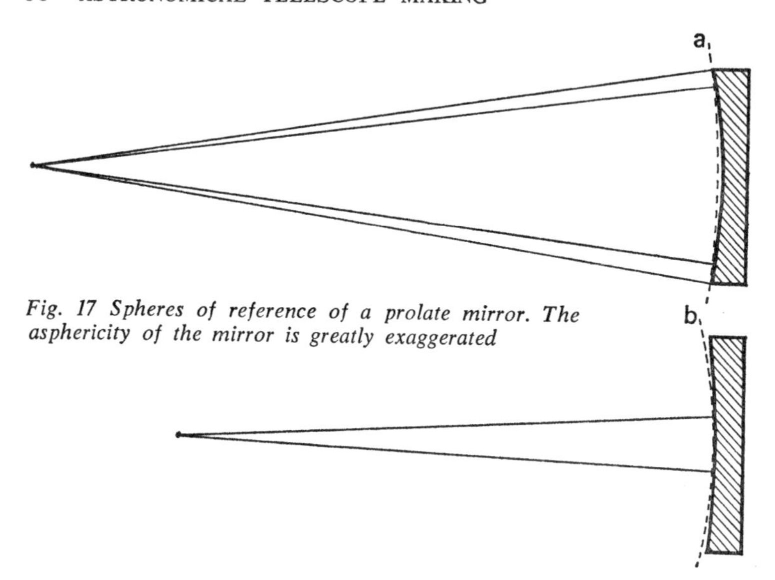

Fig. 17 Spheres of reference of a prolate mirror. The asphericity of the mirror is greatly exaggerated

a shorter radius of curvature than the margin, which appears convex, with the shadow passing across this region in the same direction as the knife-edge. Neither setting disputes the fact that the radius of curvature of the centre is shorter than that of the edge, but the mirror appears to be of a totally different shape in the two cases! Understandably, the beginner is puzzled as to what to do for the best.

Let us look at the position in another way, and consider the shape of the surface not in terms of the true concavity of the mirror, but in relation to the apparent flatness of a spherical surface under the Foucault test. An even-darkening surface would therefore be represented by Fig. 18a. The first case in the example above, with 'flat' margin and 'concave' centre, would then be represented by the cross-section in Fig. 18b, while the case with 'convex' margin and 'flat' centre would appear as in Fig. 18c. Both have the same relative shape, but have been 'bent' in different ways, and this bending is the result of our choosing different reference spheres for the two interpretations. Full understanding of the way in which a surface changes its apparent shape according to the

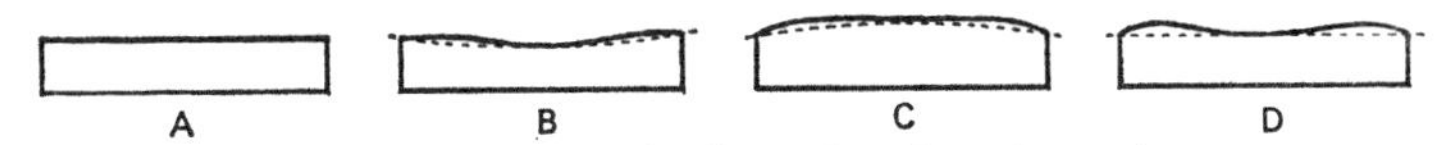

Fig 18 Apparent cross-sections of spherical and prolate mirrors

position of the knife-edge is the key to successful diagnosis in the Foucault test.

The 'best-fit' sphere

So far, we have established that the mirror has a prolate surface. But how actually does it differ from a sphere? A sphere, in terms of the Foucault test, is a shape of constant radius of curvature, and can be represented in cross-section either by a straight line (when the knife-edge is at the centre of curvature), or by smooth convex or concave lines when the knife-edge is inside or outside the centre of curvature respectively. In other words, there is no change of curve across the apparent cross-section. So we can draw, in dotted lines, the 'best-fit' smooth curves as representing the nearest spherical surface to the existing prolate one. In Fig. 18b the curve is concave, and in 18c it is convex; both represent the same curve as seen from different knife-edge positions. The difference, as can be seen, is that a 'hump' exists, which with normal prolate mirrors lies about 70 per cent of the way from the centre to the margin of the mirror.

Finally (Fig. 18d), we can flatten out the best-fit curve into a straight line and obtain the normal cross-sectional representation of a prolate surface, with both centre and edge having the same height. The hump is still at the 70 per cent zone, and this appearance corresponds to the apparent cross-section when the knife-edge is placed exactly midway between the centres of curvature of centre and margin of the mirror (Plate 2). At this point the 70 per cent zone appears as a high crest, the surface descending on either side to give a concave centre and a convex margin.

We can therefore say this of a good parabolic mirror: the distance between even-darkening settings of the knife-edge for the centre and extreme margin is equal to r^2/R, while the distance between each of these settings and the even-darkening position for the 70 per cent zone is equal to half of this amount.

Test appearance of a parabolic mirror

What do we see as we make these tests? Starting with the knife-edge well inside the radii of curvature of all the zones, we bring it back, making repeated passes through the shadow, and observe the phenomena on the mirror. To begin with, the knife-edge shadow is straight, and two or three dark diffraction lines precede it; these widen and become less distinct as the centre of curvature is approached. Before they become difficult to see, observe their shape as they pass off the upper and lower margins of the mirror. If the edge of the mirror is perfect, with no turned-off (or turned-up) edge, the diffraction lines will be straight throughout; otherwise, they will bend sharply at the margins. If they turn inwards, towards the vertical diameter, the edge is turned down; the reverse otherwise. As they cross the mirror's meridian, their direction changes and their presence is particularly easy to see. These diffraction lines represent an interval of a quarter of a wave-length of light on the mirror's surface, so that the degree of turned edge, if present, can be judged. Another test for a clean edge is the presence of a bright, narrow line of light around the mirror as the knife-edge passes across near the centre of curvature. This is caused by diffraction of light at the margin. If no bright line is present, the extreme edge cannot be sharp, but this is a highly sensitive test and a mirror showing no diffraction line, or only a faint one, may still perform well.

As the knife-edge approaches the inner centre of curvature, the middle of the shadow begins to bulge forwards, engulfing the central 30 per cent or so while the rest of the mirror is still only half-darkened. This is where the first of the human error comes into play, some workers 'taking' the centre early and others late, and it is particularly difficult to judge because the centre is so gently aspheric that the shadow is ill-defined.

The best course is first of all to go definitely past the right point, so that the region begins to darken first on the side opposite to that of the knife-edge; then go back and settle on an acceptable compromise before making a mark. As the knife-edge recedes further, the crescentic shadow on the opposite side expands rapidly outwards, reaching further and further towards the margin before the entire mirror darkens. Meanwhile, the marginal shadow on the knife-edge side narrows. Before setting up the mirror, the 70 per cent position was marked on the glass with two dabs of a felt-tip pen on either side of the horizontal diameter, and the next task is to decide the knife-edge position at which the extreme outer edge of the inner shadow and the inner edge of the outer shadow reach their respective marks simultaneously. This is the well-known '70 per cent position'. Another mark is made. Note how, in this position, the outer shadow is noticeably darker than the inner one, because the slope of an aspheric surface is steeper towards the margin; this can be seen from the cross-sections in Fig. 18.

As the knife-edge recedes still further, so the crests of the shadow continue to move outwards; and we now have the problem of deciding just when the extreme outer edge of the mirror darkens evenly. This is made more difficult by the fact that the knife-edge side of the shadow, which is now extremely narrow and only appears near the end of the knife-edge travel, appears relatively dark because of the contrast with the brightness of the mirror further in, whereas the opposite margin appears relatively bright against the adjacent dark shadow of the intermediate zone. The best way of minimising this effect is to observe the upper marginal quadrant of the mirror and to establish the point at which it darkens evenly. Go definitely past the correct position, and then back, achieving a mean setting as done for the central zone. Make another mark. If the mirror and the observer's judgment are both true, the three lines will be evenly spaced and will correspond to the formula distances. It would, however, be most advisable to take three or four different sets of readings, for only in this way can it be seen if consistent judgments of the shadow are being made. (See Plate 5.)

Over- and under-correction

Of course, the mirror may not be a good parabola. Assuming, for the moment, that turned-edge and zonal irregularities are absent, the curve may still be prolate but faulty. The most obvious reason is that the overall aberration is wrong, being either too small (mirror under-corrected, with a prolate ellipsoid shape) or too large (mirror over-corrected, or hyperbolic). In the first case, the difference between the extreme radii of curvature must be increased, either by deepening the centre or flattening the edge (or both), and in the second case the difference must be reduced by lowering the 70 per cent zone and making the mirror more spherical.

Irregular asphericity

It is also possible for the aberration to be correct overall, but with the reading for the 70 per cent zone not coming midway between the marks. In other words the shape, rather than the degree, of asphericity is wrong. This commonly occurs. Over-enthusiastic asphering can often produce a cross-section like that in Fig. 19a, which amounts to a spherical mirror with a hole in the centre (see Plate 4). In this case, the crest of the shadows will be too far in, with perhaps no 70 per cent reading possible at all because the outer part of the mirror looks flat. It is also possible for the crest to be too far out, with a shallow centre and a steeply-curved margin, as in Fig. 19b. In both cases the crests must be moved to their rightful position by local polishing on one side.

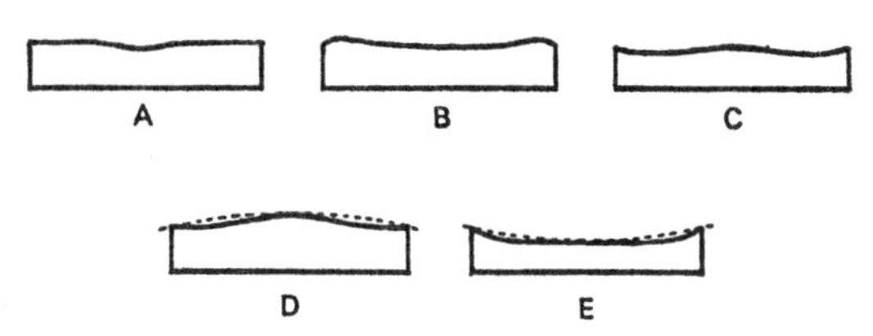

Fig. 19 Apparent cross-sections of irregular and oblate mirrors

An oblate surface

Finally, the test may show an oblate rather than a prolate surface. The cross-section of this, referred to the best-fit sphere, is as in Fig. 19c, with a high centre and edge, but few opticians would treat it in this way since two separate regions would have to be polished down. In most cases it would be 'bent' to form Fig. 19d, with a high centre that could be polished down to bring the surface spherical, with further polishing making it prolate. Other workers might instead turn it into Fig. 19e, and work the mirror face-up, with a full-size polisher, to bring down the edge.

Take heart!

By now, the reader with his humble 15cm disc may be wondering what on earth he has let himself in for, and whether mirror-making involves the mental gyrations demanded by those cosmologists who turn the universe inside-out at the flick of a constant in one of their equations. He may even have decided to abandon the whole project and send off for a finished mirror before he has a brain storm! It should, therefore, be added straight away that the mirror being described could be made perfectly adequately knowing nothing more than the right value of r^2/R and the basic elements of the Foucault test. Indeed, most first mirrors are made in this way. If the polishing has gone well, and the mirror is virtually zone-free and spherical, it may take only a few minutes to achieve an approximately parabolic shape that will give excellent results, and the worker can fling 70 per cent zones, diffraction edges and the rest to the winds. Insight is needed only when problems arise. The choice of a 15cm f/10 mirror was made deliberately to reduce the figuring problems, but the hope must be that the worker will graduate to larger mirrors with more steeply aspheric surfaces, and then he will need all the information given in this book – and more.

Removal of zones

Short-period errors, known generally as 'zones', will not arise if the precautions advised in the previous chapter were observed. The typical appearance of raised and depressed zones are shown in Plate 4, although they may not be so easy to detect if they are superimposed on aspheric shadows. By advancing or retracting the knife-edge until they darken evenly across their width, their radius of curvature can be measured and their depth calculated. This is often of interest, although it makes little difference to the type of remedial action required. Suppose, to take an example, that a depressed zone requires an advance of the knife-edge of 10mm to reach its centre of curvature, and that the measured width of the zone is 20mm, the general radius of curvature of the mirror being 300cm; the sagittal difference between equal slices of the general curve of the mirror and of the zone tells us the depression. For a 20mm slice of the average radius of curvature of the mirror, we have

$$S_m = \frac{r^2}{2R} = \frac{1^2}{600} = 0.016667\text{mm}$$

For the zone, R=299.5cm (the difference of radius of curve ture being half the knife-edge movement), and hence

$$S_z = \frac{1^2}{599} = 0.016694\text{mm}$$

The difference, 0.000027mm, corresponds to a depression of about 1/20th of the wavelength of yellow light! Such a zone, though easily seen and measured, will have a barely detectable effect on the focused image in the telescope; however, no optician of any pretensions would dream of leaving so glaring a defect on his mirror. This indicates how sensitive the Foucault test is to short-period errors; a good eye, under ideal conditions, could probably detect zones 1/100th of a wavelength of light deep. The narrower the zone, the more easy it is to spot.

Zones, as we have seen, are caused by bad contact. Raised zones, due to one or more defective facets concentrated at a certain radius, are more common. The best treatment is to re-cut the facets and sub-facets with good, wide channels, re-press most thoroughly, and polish with somewhat longer strokes and a certain amount of 'W' to distribute the wear better. Low zones, for which the treatment is the same, improve more slowly, for the entire surface of the mirror must be brought down to the level of the zone itself. Small polishers can also be used to improve stubborn zones and to prepare them for the full-sized polisher treatment.

Turned-down edge

Anyone reading through the late-19th century literature on mirror-making will encounter fearsome descriptions of this defect. Opticians at that time treated turned-edge as the incurable work of the devil, and advised grinding wide bevels or painting out the edges of mirrors so afflicted! This was due to lack of comprehension of the nature of the problem. We have already seen that face-down polishing with long strokes is quite likely to produce a poor edge, and that face-up work keeps the edge crisp.

Turn-down can affect just the extreme edge, when the diffraction line will turn dull or invisible, or the diffraction lines before the knife-edge will turn off at their very extremities, or it can be wide enough to have a measurable radius of curvature, appearing as a persistent dark line down the margin or the mirror at the knife-edge side. In this latter case its width must be several millimetres, and it is more properly termed a 'rolled-back' edge; its radius of curvature may be long by several centimetres. A severe rolled-back edge can take a long time to remove, and remedial work must start as soon as it is noticed. Many first-time amateur mirrors, thought by their fond makers to be parabolic, are little more than a sphere with a rolled-back edge; the writer suspects that his was, yet he still remembers the thrill of observing the lunar craters with this poor effort almost twenty years ago!

A turned-up edge is never a serious problem, and the writer has rarely encountered it. It may occur if very short strokes are used, in which case the remedy is obvious – lengthen them. It manifests itself either as a brilliant diffraction edge that is strikingly obvious immediately the reflection of the pinhole is picked up, or as a reversed turn of the diffraction lines before the knife-edge.

A wide rolled-back edge is a serious mirror fault, as it affects a relatively large area of the optical surface and will produce a noticeable haze around the focused image. It must be completely cured before any asphering work commences. Face-up work, with a moderate stroke, is advised. Severe cases may be submitted to the small-polisher technique advised below for treating hyperbolic mirrors.

Deepening an oblate spheroid

The classical conditions for the production of an oblate curve are face-down work on a hard, well-contacted polisher. Maximum polishing action then occurs around the zone of overhang at the end of each stroke, thus shortening the radius of curvature of the marginal regions and effectively throwing up the centre. Softer polishers tend to slump around their margins, and this digging in does not then occur.

The treatment depends to a large extent on where the crest of the shadow occurs. If it is near the 80 per cent or 85 per cent position, we have effectively a high and extensive central hill or – bringing the knife-edge back – a relatively narrow turned-up margin. The most rapid way of eliminating this defect will be to work on the narrow margin, by polishing with moderate strokes on the face-up mirror and knocking the margin down. A 70 per cent or narrower crest indicates direct work on the centre, as described below.

Reducing a hyperboloid

'Hyperbolic with a turned-down edge' describes the most common condition of a mirror polished face-down with long strokes and left to find its own way until the Foucalt test is belatedly brought to bear. A soft polisher, sinking out of contact at the margins and not re-pressed sufficiently often (if at all), will almost certainly produce a deep prolate curve.

It is sometimes said that a long-focus paraboloid, such as an f/10 mirror, is more difficult to produce than an f/8 or even f/6 mirror of the same diameter. If there is any truth at all in this, it can apply only to those workers who habitually produce prolate surfaces on their optics. But certainly the point is worth considering, for what is hyperbolic to an f/10 mirror may be only marginally prolate to an f/6. The sagittal difference between sphere and parabola for our 15cm f/10 mirror is only 0.00015mm; for a 15cm f/6 it would be some five times as much*. So the optician faced with the latter might well deliberately use long stroke to get the surface aspherising along the right lines – as long as he looked after the edge. The f/10 worker cannot afford to do any such thing, for his sphere and parabola correspond so closely; he must remain within a millimetre or two of a sphere the whole time.

Following the general rule that long strokes produce prolate shapes, we might try to reduce the hyperbola by using very short strokes. The trouble with this method is that, in pressing the polisher, we tend to impress the hyperbolic form on it also, so that the curve reduces only slowly. This may be why the old generation of opticians called the hyperbola 'fatal', and floundered miserably in its depths. Full-polisher treatment, unless the over-correction is by perhaps 50 per cent or less, can be slow and uncertain. On the other hand, it should certainly be tried first and given 30 minutes or so.

Small-polisher work

The most direct way, and the one that is sure to give the quickest results, is to tackle the high zone with a local polisher. We have already seen, in Fig. 18d, that a prolate curve can be

*The sagittal difference, δS, between the sphere and the paraboloid is given approximately by the formula $\frac{D}{50}\left(\frac{10}{f/}\right)^3$, where D is the aperture of the mirror in cm and f/ is the focal ratio. The units are wavelengths of yellow light. See the table on page 167.

considered as having a surface in the form of a double curve (American writers refer to it as the 'doughnut'). A hyperbolic mirror has too extreme a curve, and its crest is too high; the logical answer is to polish it down by the appropriate amount.

Small-polisher work is the most efficient way of amending the curve on most forms of optical surface, because it can be directed precisely at the glass that needs to be removed. All professional opticians figure in this way, for full-size polisher work has the grave disadvantage that it cannot be completely controlled. All that can be said is that such-and-such a stroke will *tend* to give a certain result. If conditions are precisely similar, then the same stroke must clearly do the same thing at all times; the problem with optical work is that there are too many variable factors to allow us to speak in more than generalities. A slight temperature change, or a fresh system of facets, or wetter or drier rouge, can all change the result even though the nominal strokes used are similar. If we are polishing away at the whole surface while trying to exert local changes, these imponderables can upset the best-laid plans. Local working with a small polisher is much more repeatable; but it is also drastic, because there is no longer the automatic blending tendency of the full-size polisher. This means that it must be used in sparing bouts if severe zones are not to be formed. The 'zoning' reputation of small polishers has undoubtedly discouraged amateurs from using them, but steeply aspheric surfaces cannot be made reliably in any other way, and it is a technique which must be mastered by the serious optician. The sooner it is attempted, the better!

To be of any use, a small polisher must be *small enough.* A $\frac{1}{2}$- or $\frac{2}{3}$-diameter polisher will be unable to do anything very much except produce an oblate spheroid where its edge works on the intermediate mirror zone. A $\frac{1}{3}$-diameter polisher, with its pitch surface trimmed to a rough star shape, will be able to handle the high zone of a hyperboloid and also polish down the centre of an oblate spheroid. A slightly larger diameter, say $\frac{3}{5}$, would be ideal for parabolishing an essentially spherical mirror.

Small polishers can be made from thick plywood, filed or turned circular, and well varnished to make it waterproof. Some special soft (15-second) pitch can be made up in a small tin; a 3mm layer is poured on the disc and pressed to curve on a splash of rouge on the mirror's surface. Cut facets in it when it has cooled. They will quickly fill in during working, and should be renewed regularly. Some workers leave a full, circular working surface, but trimming the margin to a star-shape will help somewhat in blending in the wear, particularly in inexperienced hands.

Reference again to Fig. 18d will show that the glass is not heaped up around the crest symmetrically; the outer slope is noticeably sharper. The centre of the polisher should therefore follow the path indicated in Fig. 20a, so that the maximum wear occurs just outside the crest rather than on it. The position of the crest itself can be marked on the back of the mirror with a felt-tip pen after it has been marked on the front surface during knife-edge testing. Five or six 'scallops' should take the polisher round the mirror once in an interval of 2–3 seconds. Meanwhile, the operator walks round his mirror as usual, and rotates the small polisher in his fingers. Test after not more than five minutes' working, allowing the mirror at least twice as long to cool down. If the asphericity is less, and the crest is in the same position (assuming that it was in the right position to begin with), the work is clearly going along the right lines.

Small-polisher work is also the most direct way of treating irregular aspheric surfaces, where the crests are in the wrong place. A hole in the centre, as in Fig. 19a, will respond to work just outside its crest (Fig. 20b), while a wide, flat centre (Fig. 19b) requires the oblating stroke just described for a hyperboloid, but with the passes occurring nearer the edge of the mirror. It may even be necessary to make a slightly smaller polisher to get the wear sufficiently outside the crest to bring it inwards (Fig. 20c). Progress when figuring with

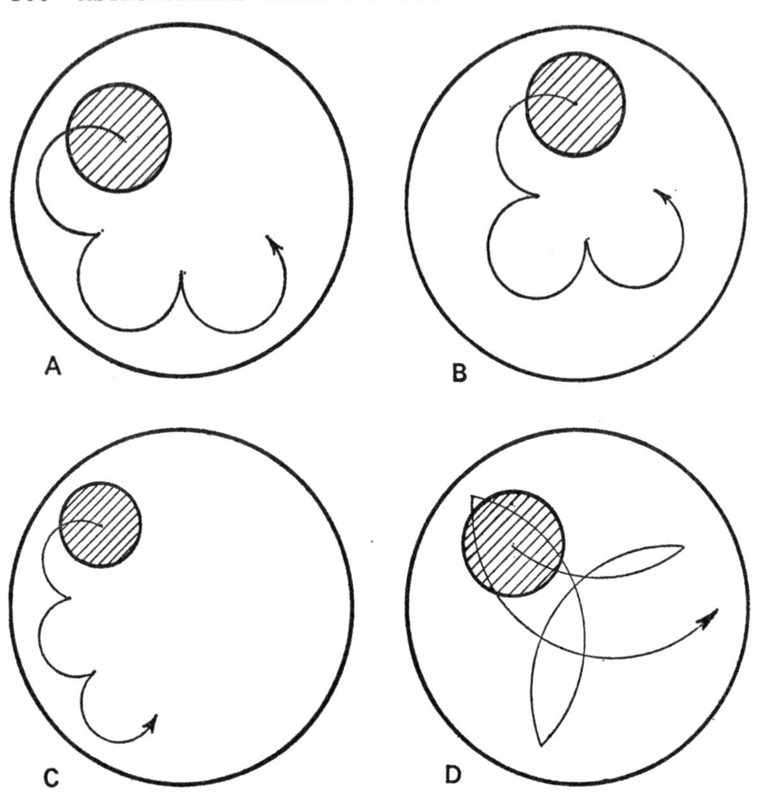

Fig. 20 Strokes using small polishers. The polisher is represented by the shaded disc

small polishers is always so rapid that it can quickly be established if the adopted stroke is the right one.

Until the worker becomes used to the method, this rapidity of action will come as a considerable surprise. Just 30 seconds of zonal work can make a detectable difference to the appearance of a mirror under the Foucault test. (This sensitivity, incidentally, is in direct proportion to the radius of curvature of the mirror being tested. For example, we have already seen that a 20mm zone 1/20th of a wavelength of light deep will produce a knife-edge movement of 10mm if the general radius of curvature is 300cm. If this radius of curvature were only 100cm, the difference of radius of curvature of such a zone would be only 3.3mm, and it would not look nearly so prominent. Nevertheless, it would have the same effect on the final image.) It may be asked why, since the area of the polisher is so small; and the answer here is that we are seeing the *true* rate of the polishing action rather than the *difference* between the rate of polishing achieved on different zones of the mirror by a full-size polisher. Thus, the large polisher will take off glass as rapidly as the small one, but from all over the mirror at almost the same rate. Figuring by full-size polisher therefore involves a good deal of wasted labour in removing glass that might just as well stay put.

Small polishers, then, must be used in sparing spells, and it must not be expected that they will be able to remove large excesses of glass in a completely zone-free manner. If, however, the mirror is between 50 per cent and 100 per cent over-corrected, small-polisher work should be expected to wear down the crest before any serious zones become noticeable; indeed, the mirror may well be acceptable. If zones do begin to appear, then re-press the full-size polisher and use normal strokes to remove them. The over-correction will then have been removed, which was the object of the exercise.

Parabolising from a sphere

In general, it is easier and more satisfactory to deepen a sphere into a paraboloid than to have to reduce a hyperboloid. A sphere is the easiest surface to monitor for zones, turned edge, etc., and changes in its shape, as figuring progresses, can be observed more easily. Therefore the beginner would be well advised to bring his mirror as near spherical as possible before thinking in terms of parabolising. Of course, a paraboloid may be achieved, by accident, during the course of the work, and a slight hyperboloid at the end of polishing can be reduced easily into the required paraboloid; but this is the luck of the game.

It is common to think in terms of parabolising a sphere as a matter of deepening the centre to produce the required aberration, and long stroke or W-stroke is advocated. If this method is adopted, it is of the utmost importance that it be confined to 5-minute spells with rigorous cold-pressing to follow; otherwise, the deforming polisher may produce a hole in the centre, with a turned-down edge as well. This makes the method a slow and uncertain one, at least when trying to parabolise an f/8 or faster mirror, although it may more often be successful at f/10, when the asphericity is only slight.

The opposite approach, in which the mirror is turned face-up and a long stroke is used to flatten off the margins, has the advantage that the edge is kept crisp. A half-diameter stroke, with the polisher overhanging by 3–4cm at the extremity of the 'W', can be maintained for 10–15 minutes at a time, after which an aspheric shape of some sort will have appeared. The important thing is to see where the mid-way knife-edge crest appears; if it is near the 70 per cent zone, well and good, and an f/10 mirror will probably be within the permissible limits of accuracy. Very little polishing is required to parabolise an f/10 sphere, provided the sphere is a smooth one and the crests appear in the right place.

Less hit-and-miss is the use of a small polisher of about $\frac{2}{5}$ diameter on the centre of the face-up mirror, since the crest of the shadow can more easily be moved about if it is in the wrong place. Of course, there is nothing to stop the use of a small polisher in conjunction with a full-size one as described above, and this has the advantage of keeping incipient zones under control. Small-polisher parabolising, however, has much to recommend it, particularly in the shorter focal ratios. The object is to remove most glass from the centre of the disc, but this will produce a violent hole if nothing is done about blending in the edge. We therefore polish in long swathes (Fig. 20d) that pass over or near the centre of the mirror, but also allow the polisher to touch the margins. A five-minute spell of this work may well have produced a good parabolic shape. If the crests are wrong, they can be treated with the same polisher, or a slightly smaller one, as described above.

Surface accuracy required

Since a parabola is an effectively non-existent frontier between two families of curves, the prolate ellipses and the hyperbolas – both of which have infinite variations – it is reasonable to ask how closely our mirror's surface must approach to this unattainable ideal to be considered satisfactory.

This subject was discussed exhaustively back in 1878 by Lord Rayleigh (1842–1919), the famous physicist, who investigated the quality of the star images produced by telescope objectives of different standards. If we consider the light coming from a star as a series of wavefronts being radiated into space, then the task of the telescope objective is to condense these waves to form a point image. To do this properly, the wavefront produced by the objective must be a spherical one (Fig. 21). An objective afflicted with spherical aberration will not produce a spherical wavefront, and the waves cannot, therefore, converge to form a unique image. Rayleigh's experiments indicated that the definition afforded by a telescope will be effectively perfect if the converging wavefronts are spherical to within $\pm\frac{1}{8}$ of their wavelength, and this per-

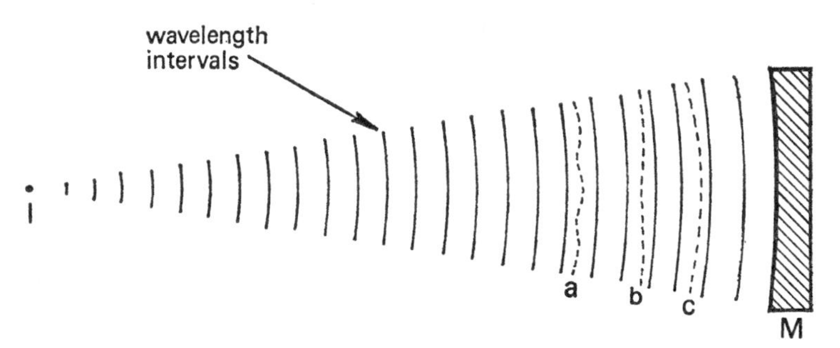

Fig. 21 Wavefront errors. A schematic representation of the spherical wavefronts that should be produced by a perfect mirror (M) to form a perfect image (I). The Rayleigh criterion allows wavefront errors of $\frac{1}{4}$ wave in total amplitude, and three possible forms are shown: (a) irregular; (b) under-corrected; (c) over-corrected

missible error in the wavefront represents the *Rayleigh limit.* The errors on the actual surfaces of reflecting systems are represented by about double the error in the resultant wavefront, so that a mirror figured to the Rayleigh limit will be parabolic to within $\pm\frac{1}{16}$ of a wave – which is another way of saying that no region on its surface departs by more than $\frac{1}{8}$ of a wave from the best-fit paraboloid, all departures being measured in the same direction.

The total aberration on a 15cm f/10 mirror is only 0.3 of a wavelength of yellow light (which is normally taken as the standard, because it is the colour to which the normal eye is most sensitive), and this is represented by a knife-edge movement of 1.9mm. Therefore, if the mirror is zone-free, and its measured aberration lies between 1.2mm and 2.6mm, we know that the correction falls within the limits of the Rayleigh requirement (for an f/8 mirror, the limits would be 1.9mm and 2.8mm). This is, relatively speaking, a huge range. If we can hold the limits to within half these amounts, the mirror will be so good that its performance, at all times, will be limited not by its own quality but by that of the diagonal, or eyepiece, or atmosphere – or observer! This is a much preferable state of affairs, for the errors in an '$\frac{1}{8}$th-wave' mirror, while small, are perfectly appreciable to an experienced eye.

The diffraction image

The image of a star formed by a perfect telescope, as represented in Fig. 1, consists of an 'Airy disc' surrounded by several diffraction rings; of the total light focused by the objective 85 per cent is concentrated in the disc, and the remaining 15 per cent is distributed in the rings. The effect of declining surface quality is to take light away from the disc and add it to the rings. If the mirror is over- or under-corrected by $\frac{1}{8}$ of a wavelength, the proportion of light in the central disc drops to 68 per cent – this fulfils Rayleigh's criterion. If the error is as much as $\frac{1}{4}$ of a wave, only 40 per cent disc illumination is achieved. Beyond this quality the term 'disc and rings' ceases to have any effective meaning.

and the telescope will be defining far below its potential; a star image will be represented by a blurred patch several times its theoretical size. If we consider the surface of an extended object, such as a planet or the moon, as being made up of discrete 'starlike' points, then its image in the telescope will similarly be compiled from a great number of disc-and-rings components, and the contrast in the features being viewed will be dependent on the amount of light concentrated in the discs. Putting excess light into the rings will severely lower the contrast, and hence the amount of fine detail visible. Therefore, first-class optics are particularly necessary for *planetary* work, and the last fraction of a wavelength of precision in the figuring will be rewarded by a distinctly improved 'bite' in the delineation of planetary markings. Stars, being seen against a dark sky, are objects of much higher contrast, and suffer less from augmented ring brightness unless we are seeking a faint star in the immediate vicinity of a bright one.

Even a *spherical* 15cm f/10 mirror will be almost within the marginal $\frac{1}{4}$-wave tolerance, and it must be confessed that probably most amateurs' first attempts at the usual 15cm f/8 design fall outside this limit because of the difficulty of testing and treating the steeper asphericity. A 15cm f/12 mirror could be left spherical and would almost achieve the Rayleigh limit.

Other tests

The Foucault shadows on an f/10 mirror are the merest 'breaths' of shade, and vibrations due to people walking about in the house, or the swirl of air waves across the mirror, can make them difficult to judge, particularly when it comes to measuring zones. An eyepiece test is less affected by these interfering factors and is often useful; and, of course, it is the classical test for a spherical mirror at the centre of curvature, when the image of the illuminated pinhole should be seen perfectly defined. The eyepiece can be hand-held, but is more conveniently mounted on a stand with some height adjustment to aid in centring the image. A focal length of about 20mm is suitable. A perfectly *spherical* mirror will give the appearance shown in Plate 6, with the intra- and extra-focal images looking identical and somewhat more brightly illuminated at the margins, due to the edge diffraction around the mirror. If the pinhole is small enough, the expanded discs will be seen to consist of a series of concentric diffraction rings. Any short-period error on the mirror will now stand out. If the zone is raised, and hence focuses long, the corresponding part of the expanded disc will appear relatively bright outside focus, and faint inside; a depressed zone will give the reverse effect.

Of course, an aspheric mirror will no longer give similar intra- and extra-focal expanded discs when viewed at the centre of curvature, any more than a spherical mirror will do so when focused on a star. If the mirror is prolate, with the centre having a shorter radius of curvature, the centre of the inner expanded disc will be relatively bright, as will the edge of the outer disc. These appearances are shown in the plate. An oblate spheroid will give the opposite effect. Despite these anomalous appearances, zonal irregularities are still detectable. In fact, an experienced eye can identify them more clearly than when using the knife-edge.

Telescope testing must be the ultimate appeal of the optician. The Foucault test, for all its sensitivity, does not give a direct judgment on the mirror's practical performance, and not until he has acquired much skill can the optician be confident that he fully understands what the shadows mean. Once the mirror appears to be finished, a star-test should be made. If the tube is not yet completed, a rough wooden rig can be made up to hold the mirror and diagonal in their respective positions; the eyepiece can be held in the hand. Then locate a bright star, such as Sirius or Vega, in the 'telescope' (a friend to aim the affair and hold it steady, will prove invaluable). Bare glass reflects about 4 per cent of the light falling on it, so, provided the diagonal is coated, the star image will be quite bright enough for its features to be seen. The critical test is how similar the expanded discs appear when observed about 3mm from the focus with a 10mm focal length eyepiece (x150). Plate 6 shows various appearances. The rule to remember is, as before, that *high* regions appear *bright* in the *extra*-focal disc, and vice versa. If a proper mounting is available, it will be possible to knife-edge the image, and here, of course, a truly parabolic mirror will darken evenly all over – a null test – although an uncoated mirror will probably not give a bright enough reflection for the zonal shadows, if any, to be seen.

Remembering that this is a highly critical test, do not worry about any differences that are too slight to be certain about. But if the margins of the two discs are glaringly different, one being bright and well-defined, the other fading away with no definite boundary, then something is amiss. It may be rolled-back edge, or under- or over-correction. Knowing that the condition is present, subsequent appeal to the Foucault test will enable the fault to be located and treated.

Mirror coatings

For visual work, fresh silver reflects more light than any other single coating (about 94 per cent of yellow light), but it tarnishes so rapidly in air that after a few weeks the reflectivity may have dropped to 85 per cent, and it will continue to deteriorate unless very well looked after. Silvering can, however, be done at home using easily-available chemicals (instructions and references for this are given in the *Amateur Astronomer's Handbook*, Appendix C). Aluminium. with an initial visual reflectivity of about 87 per cent, is much more resistant to tarnish and maintains its brightness for many years if the mirror was well-coated to start with and is periodically washed with pure industrial detergent, such as Teepol. Aluminising is performed by evaporation of the metal in a vacuum chamber, and firms performing this service are listed in Appendix A.

CHAPTER SEVEN

Tubes and Mountings for Reflecting Telescopes

It would be fair to say that no two amateur-built telescopes have ever been alike. There are so many ways of going about the task, both in design and materials, that the finished product will be as unique as the maker himself.

However, in Chapter Three we laid down some principles of design concerning the beginner's reflector. These can be extended and summarised as follows.

1. Simplicity of construction
2. Rigidity
3. Easy pointing and following facility
4. Permanence of optical alignment
5. Relative portability

Simplicity of construction

Wood is not cheap to buy, but it is easily worked and assembled, and there may be enough lying about the house and workshop to provide useful material.

Rigidity

Almost everything that can be said about any particular mounting revolves around this point. Rigidity of mounting is more important than aperture and – up to a point – even more important than optical quality; for, if the image dances in the eyepiece, it matters not if it is good or bad! More telescopes have been spoiled by shaky mountings than by any other cause.

A powerful telescope can be considered as a long lever magnifying its own tremors by upwards of 300 times. If the upper end of our 15cm f/10 telescope tube shakes by only 1mm, the image of the planet Jupiter will vibrate by three times its own diameter! To constrain such a tube rigidly within these limits would be impossible without a colossal stand; in practice, we require that the tube should not shake excessively when the eyepiece is focused or the slow-motions are used, and that the vibration should damp down quickly once the interference is withdrawn. What constitutes 'excessive' will be obvious from experience.

Easy pointing and following

The amount of sky shown by a telescope with a high-power eyepiece may be as little as $\frac{1}{3}$ of the apparent diameter of the moon, which is the same as 1/9 of its area. A wide-field finder telescope, with cross-hairs marking the centre of the telescopic field, is therefore essential for the rapid acquisition of objects. The telescope itself must be equipped with coarse and fine motions to allow the tube to be pointed accurately and to enable the observer to compensate for the earth's axial spin, which can move an object right across a high-power field of view in 30 seconds, or even less. These motions must be quick-acting but delicate if prolonged telescopic observation with an altazimuth mount is to be feasible.

Permanence of optical alignment

The field of view of critical definition of a paraboloidal mirror is relatively small; in the case of a 15cm f/10 telescope, the Airy disc has doubled in size, due to the off-axis effects of coma and astigmatism (but mostly coma), at a distance of only $\frac{1}{4}°$ or about 6mm from the optical axis*. The definition of an f/8 mirror falls off about twice as severely. This means that the mirror must be squared-on sufficiently accurately to bring its optical axis to well within these limits of the centre of the drawtube – and its cell must also be able to maintain

*Coma does not affect spherical mirrors, so under-correcting a mirror improves its off-axis performance, although at the expense of axial definition.

this positioning at all altitudes of the tube. The tolerance reduces swiftly as the focal length of the mirror shortens, the angular field limit of good definition reducing as the square of the focal ratio: another mark in favour of an f/10 system.

Relative portability

Given that none of the factors mentioned above is sacrificed, we have already seen that portability is, to the average observer, a useful feature of his telescope.

A proposed telescope mounting

The following notes are not intended to constitute a definitive description of how to make a tube and stand, but they indicate one way of doing the job, and variations are either outlined or left to suggest themselves. Figs. 22 and 23 show how the component parts are assembled, and the frontispiece shows a complete telescope made from these directions. The description is for a 15cm f/10 telescope, but an f/8 telescope will differ only in the length of the tube. Most dimensions are approximate, and can be modified to suit the material available; the exposition is one of principle rather than of practice!

The tube

This is of square section. The four corner pieces are 160cm long, of 4 x 4cm wood, and form a long box with sides 20cm wide. These sides are closed in for about 30cm at each end, using 5-ply or other suitable material. The frame is reinforced at its upper end, and halfway along its length, with 7.5 x 2.5cm strips. The lower side also carries a stretcher of 7.5 x 5cm wood, 40cm long, to which hinges for vertical motion will later be fixed. The lower end of the tube is blocked in with thick (20mm) plywood drilled with three appropriate holes on a central 12cm circle to take the mirror cell adjusting screws. A concentric 8cm hole is cut inside this circle to allow access to the inner adjusting nuts.

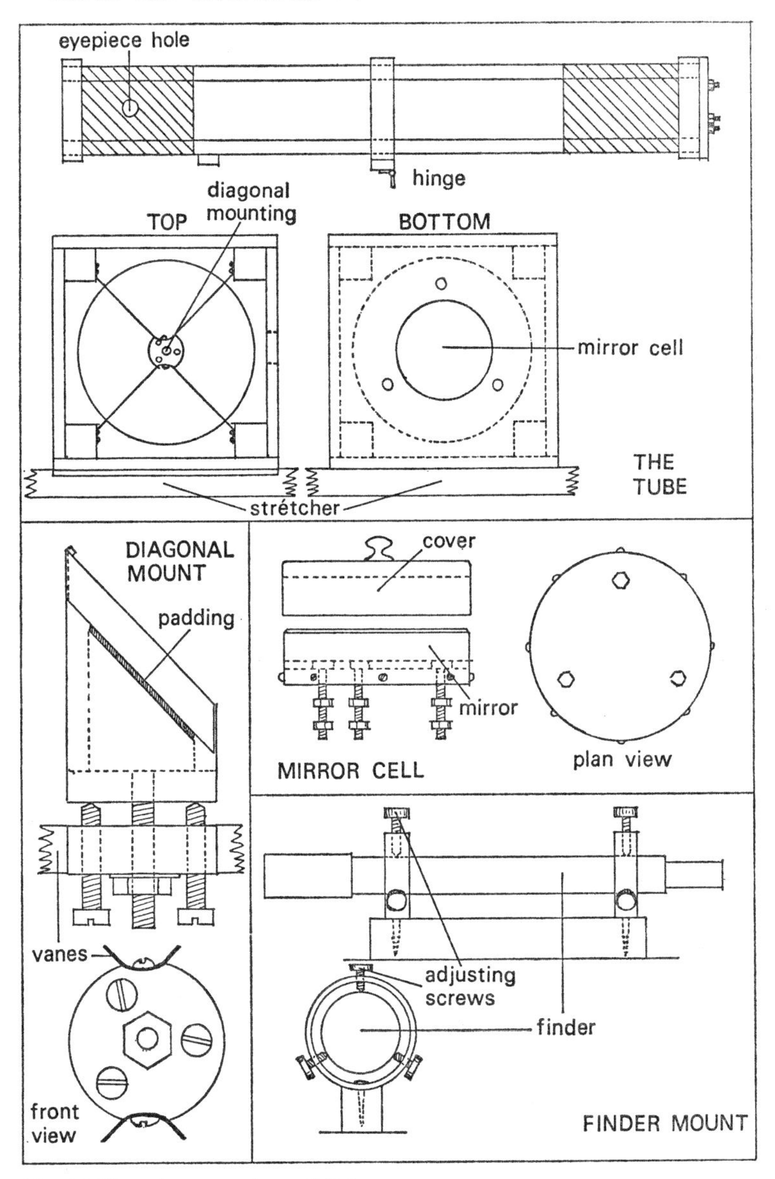

Fig. 22 The telescope tube and fittings

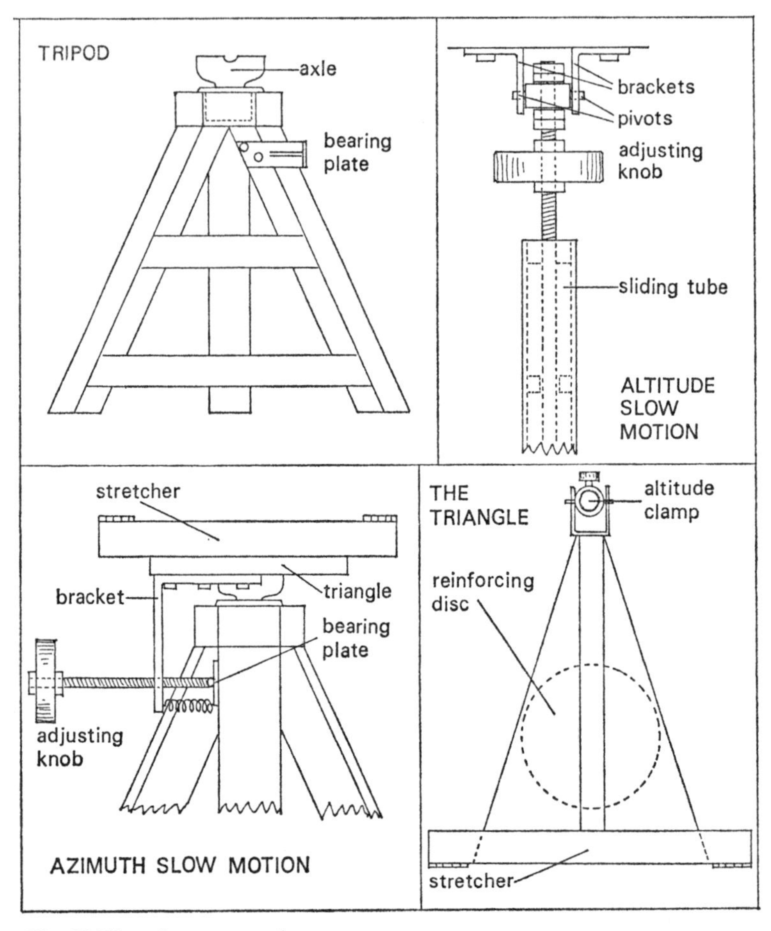

Fig. 23 The telescope stand

The mirror cell

Saw out and file true a thick plywood disc about 5mm wider than the mirror. Drill three holes on a 12cm circle to take long screws or lengths of screwed rod that will pass through the base of the tube and leave room on both sides of the base for adjusting nuts. The ends of the screws inside the cell are left slightly proud to act as a 3-point support for the mirror disc.

Stand the cell on the bench and set the mirror on it. Run

some strips of felt or baize around its edge until the total diameter matches that of the cell base, and screw a strip of flexible light-alloy sheet to the rim of the cell so that it comes up to the bevel of the mirror and holds it firmly within its felt lining. The mirror is now safely in its permanent cell housing.

Make a cover for the mirror by taking another plywood disc of suitable size and running a strip of light-alloy sheet round its edge so that it can be fitted down over the cell. Before placing the cover over the mirror, a thick cardboard disc faced on one side with a piece of linen or other soft cloth and lightly wadded with cottonwool should be laid on the optical surface. This, when held down with the wooden cover, will provide perfect protection. Do not allow this soft liner to become damp; it is a good idea to leave it indoors in a warm place during an observing session.

The diagonal fitting

The best holder for the elliptical flat is a short length of tube into which it will just fit, cut off at one end at an angle of 45°. A narrow piece of flat metal across the extreme end of the ellipse so formed will hold the diagonal in place when it is subjected to slight pressure by a 45° block, faced with felt, attached to the inside of the wooden or metal disc used to form the base of the holder. A piece of screwed rod extends from the centre of this base and passes loosely through a hole drilled through the centre of a mounting disc. Adjustment of the inclination of the diagonal is provided by three screws passing through this disc and bearing against the base of the holder, the slack being taken up by tightening a nut on the central rod. In this way, the diagonal can also be set exactly opposite the eyepiece tube, should the mounting disc have been positioned somewhat inaccurately. This mounting disc can be of thick (15mm) plywood, with the screw holes drilled slightly undersize, the screws making their own thread as they are passed through. The diameter of the disc should be no more than that of the holder, as we wish to obstruct the minimum amount of light – a tube of about 35mm outside diameter should be able to accept the diagonal.

If no tube is available, the diagonal can be mounted on an inclined elliptical disc which is fastened to the baseplate, or a sawn-off wooden rod can be used. The diagonal can be held by clips, or simply by adhesive. If adhesive is used, it should be of the plastic sort that never sets hard; epoxy resin types can exert enough force to twist even relatively thick plates of glass.

A point often overlooked is that the diagonal needs to be looked after just as carefully as the main mirror, and it should always be covered when the telescope is not in use. If mounted in a tube, a larger tube in the form of a cap can be slid over it; otherwise, some loose-fitting cover must be made. The surface can be protected with a small wadded pad.

The mounting disc must now be supported in the centre of the tube and at the correct distance from the mirror. A single arm extending from the upper or lower side of the tube can be used, but a more robust way is to take two strips of the light-allow sheet already mentioned, each about 30cm long and as wide as the thickness of the mounting disc. These are then secured to opposite edges of the disc at points halfway along their length, and bent to form four vanes which are screwed to the inside of the tube formers. Care should be taken to see that the plane of the vanes is parallel to the axis of the tube, so that only the thin edge outline is presented to the mirror. The diagonal unit is best fitted after the eyepiece hole has been cut.

The focusing unit

A simple sliding unit, consisting of two nesting brass tubes, can be used, but a rack-and-pinion system, which gives smooth motion at the turn of a knob, is much to be preferred. Helical or screw-type focusing systems are considerably cheaper, but some people object to the fact that the eyepiece rotates while being focused. A more serious disadvantage of this type of system is that most helical units have a focusing range of only a couple of centimetres, which may not be enough to accommodate all the eyepieces in the observer's box. Rack-and-pinion systems have a sliding drawtube in addition to the racked tube, allowing a total range of 10cm or more.

In this country there are two standard eyepiece mounts. The original standard was the so-called R.A.S. threaded type, in which the focusing mount carries a female 1¼in. x 16 tpi fitting. This allows the use of eyepieces covering a linear field of up to 25mm across, which, in conjunction with an objective of 150cm focal length, represents an angular field of about 1° in diameter: adequate for enjoyable low-power views*. In recent years, eyepieces made in Japan with a 24.5mm push-in fitting have been imported in large quantities, but the maximum field of view available with these eyepieces is restricted by their smaller limiting diameter. However, 24.5mm types are now so common that every telescope should be able to accept them. The best solution is to have the focusing unit fitted with an R.A.S. thread, with an adaptor handy when push-fit eyepieces are required.

To determine the position of the focusing mount on the side of the tube, set it to its halfway extension and measure the length to its base. Then add on half the width of the tube, and subtract this from the focal length of the mirror. The hole for the mount should then be cut at this distance up the tube from the mirror's surface. The vane or arm for the diagonal can then be fitted so as to bring the centre of the diagonal precisely opposite the eyepiece tube.

The finder

This essential adjunct can be bought, complete with mounting brackets, from a supplier, but the price of complete units, even small ones with object glasses of 30mm or less, is now £15–20. An excellent finder can be made for much less using

*To obtain the focal plane scale, divide the focal length of the telescope by 57.3. This gives the linear size of the image of an object 1° across. The sun or moon is about ½° across.

ex-government equipment; or, if a damaged pair of binoculars is located, the good side can be sawn off and used. The writer has a particular liking for erect-image rather than inverting finders, the reason being that the finder is used as the next step from naked-eye location, and it helps if the finder field duplicates that seen in the sky with respect to orientation. However, most commercial finders give an inverted image. The most important feature is that the aperture should be large (preferably 50mm), and the field of view should be wide enough to show about 5° of sky, which suggests a magnification of ×8 or ×10 if an eyepiece with an apparent field of between 40° and 50° is used.

Given suitable lenses, the construction of a simple inverting finder will present no difficulty to anyone who has made the telescope described in Chapter Two. Focal lengths of about 250mm and 25mm for object glass and eyepiece will be found suitable. If a 'positive' eyepiece is used (see page 134), as will normally be the case, cross-wires can be cemented on to the field stop to mark the centre of the field for alignment purposes. These 'wires' should be thick enough to show up against the night sky, and human hair will serve admirably. Three strands, forming a small central triangle inside which the object can be located, are preferable.

The finder should be mounted near the eyepiece at the points of three screws passing through two separate rings that are about a centimetre larger than the finder tube. These rings are themselves mounted on stalks, or on a wooden block, to bring the finder sufficiently far away from the telescope tube for comfortable viewing. Do not forget to provide an adequate dewcap; it is irritating to have to keep wiping a 'blind' objective when objects are being sought in rapid succession on a damp night.

The tripod

The legs of the tripod can be made from 10 x 5cm or other heavy material; the lower ends can be cut to form blunt points where they stand on the ground. A height to the top of the tripod of 60cm will be about right, so that a total length of 75cm will be needed before each leg is trimmed to allow it to stand at an angle of about 30° to the vertical. The tops are screwed very firmly to a thick plywood sandwich. The centre of this sandwich is cut out to take the barrel of a car's hub bearing, which is bolted down with coach bolts, leaving the rotating lugs uppermost. The legs must now be cross-braced with 5 x 5cm timber near the base: each length will measure about 60cm. Further braces halfway up the legs will do no harm. The wood should be well treated with preservative, especially if it is to be left outside in all weathers, and screws should be greased before insertion to inhibit rust.

The triangle

The triangle is the item connecting the tube to the stand, and it *must* be rigid, or the mounting will never be satisfactory. The rear side, which takes the tube stretcher, is 30cm long, while the distance to the front apex is about 45cm. It should not be less than 4cm thick; a sandwich of plywood, screwed and glued, will be satisfactory enough, but a reinforcement of 5 x 5cm material along the upper meridian and along the rear where the stretcher is hinged, will help. The most play-prone part of all is the connection to the hub bearing shaft, which turns inside the barrel fitted to the stand. If the hub base is not wide enough to afford a good seating, the mounting will profit greatly if a metal disc of between 5mm and 10mm in thickness and perhaps 20cm across can be bolted to the hub base to act as a foundation for the triangle. The pivotal position is about 20cm up from the rear of the triangle.

When this is finished, the tube stretcher is linked to the rear of the triangle with two heavy-duty hinges. All that is needed now is some means of holding the telescope tube in the right position, and the instrument is ready for use.

The azimuth motion

Large changes of azimuth are accomplished by turning the

whole stand bodily. For fine adjustment, two robust metal projections are fitted approximately facing each other, one below the rear of the triangle at the observer's side, so that it is near his right hand, and the other to the upper part of the stand. A length of screwed rod passes through the first plate and its rounded end bears against the other; a strong spring, hooked through holes in both plates, tends to pull them together and keeps the system in tension. A Whitworth or U.N.C. thread on 10–15mm studding will be found suitable. If no hole-tapping facilities are available, a suitable nut can be filed roughly circular and forced into a hole cut in the triangle projection. The bearing end of the rod can be made to work in a shallow groove filed in the fixed plate; it cannot be constrained tightly because of the different angles that the plates make with each other as the telescope moves through an arc. A smoother motion will be obtained if a large ball bearing is cemented to the end of the studding. A screwing range of 5cm will supply at least an hour's fine motion, long enough for the most ardent amateur without taking a break to wind the rod back! A large, easily-gripped knob completes the unit, and this can be a plywood disc secured with nuts to the end of the studding.

The altitude motion

Take a metal tube about 20mm wide and 70cm long, and into one end fix a rounded-off nut through which about 20cm of studding, similar to that used for the azimuth motion, can pass. Make sure that the nut is square-on, or the studding will emerge at an angle to the tube and the unit may seize up; it is helpful to insert an unthreaded liner, through which the studding can only just pass, a little way down the tube. This ensures alignment. The upper part of the studding carries an adjusting knob, and at the very top it passes freely through a metal disc; a large nut with its thread filed out would be suitable. Locking nuts and spring washers hold this so that the rod can turn freely inside the disc with as little play as possible. Small holes (about 4mm) are drilled in opposite sides of this disc, and short projections are cemented or hammered into place. The ends of these projections are carried by two brackets screwed to a length of wood which is subsequently fixed to the lower side of the telescope tube.

Another pair of brackets carrying a similar disc, but this time large enough to accept the metal tube, are fitted to the front of the triangle. Thus, the telescope tube can move freely up and down as the altitude motion slides through the lower unit. A locking screw passing through the side of this disc holds the telescope at the desired altitude, and fine motion is supplied by twisting the knob on the screwed rod. This thread will be held in slight tension if the pivoting of the tube is arranged to make it somewhat bottom-heavy.

The best position for the upper bracket to be fixed to the telescope tube is determined by experiment. When this is done, the tube should be able to point at all useful angles without the telescope, or the altitude motion, fouling the stand.

Completing the mounting

If work on the tube and stand is commenced with the mirror grinding, there is every chance that the two will be finished at about the same time, and the uncoated mirror can be given a thorough star-test before being sent away for coating.

White is the best colour for the telescope tube, so that it can be made out easily in the dark. The interior flat faces can be painted dull black to reduce glancing reflections from bright celestial or terrestrial objects. All mirror covers and other detachable and miscellaneous items, such as eyepiece boxes, should be painted white so that they can more easily be found after an observing session. The telescope itself can be carried into a shed or the garage, or left outside under a waterproof sheet, and will deteriorate little if the moving parts are kept greased. Such an instrument will give service out of all proportion to the time spent in making it.

Equatorial mountings

It is not the writer's intention to discuss in detail any more elaborate mountings than the one just described. The assumption, and it seems a reasonable one, is that the beginner wants to complete an effective telescope as quickly and as cheaply as possible. Time and again the writer has heard of ambitious projects being initiated with enthusiasm, but little coming of them after the first week or two of intensive effort. It is as true in astronomical observation as anywhere else that results rest not on the sophistication of the equipment used, but on the user's enthusiasm. It is a great shame that so many books virtually discount the altazimuth telescope as little more than a toy. By doing so, they involve the innocent reader in the complications of making an equatorial – which is much harder to make well than an altazimuth – and augment the risk of his being dissatisfied with the final result. Very possibly the advocates of 'an equatorial mounting at all costs' are either mechanics rather than observers, and enjoy the work (which is fair enough), or else have been brought up on commercially-made equatorial instruments and view the altazimuth telescope as unbearably primitive. Neither viewpoint is appropriate in the context of this book.

For completeness, however, a few words must be said about equatorial mountings, against the time when our beginner has become an observer (and, we hope, an optician) of some experience, and begins to think in terms of a larger telescope with the luxury of automatic following of celestial objects.

The principle of the equatorial

The earth rotates on its axis in a west-to-east direction once a day. This means that the sun and the stars appear to cross the sky from east to west, and that the points around which they seem to revolve are marked out in the sky by the projection of the earth's axis. These points are known as the *celestial poles*. Someone standing at the earth's north pole would find the north celestial pole directly overhead, and a bright naked-eye star, called the Pole Star, happens to lie very close to the true north celestial pole. The south celestial pole, which, of course, can be seen only from the southern hemisphere, has no prominent marker star.

The principle of the equatorial mount depends on the alignment of one of its axes, the *polar axis*, with that of the earth. If this axis rotates in the same time that it takes the earth to spin, but in the opposite direction, the telescope will remain pointing in the same direction relative to the stars. It should be noted that, if a motor drive is used, its period of rotation is not 24 hours but about 23 hours 56 minutes, which is the time taken for a star to appear to revolve once around the celestial pole. This is, in fact, the true or *sidereal* period of the earth's rotation. We owe our 24-hour day to the fact that day and night are dictated by the sun, not the stars. As the earth moves round the sun, it has to spin through slightly more than one complete rotation to bring the sun back facing the same hemisphere again.

The polar axis

The first business of making an equatorial mounting concerns the polar axis, for upon the accuracy of its alignment, and of course its rigidity, depends the basic success of the telescope. The angle that the axis makes with the horizontal plane is equal to the latitude of the observer, so that in the British Isles it will be somewhere between 50° and 60°, and it will be aligned in the direction of true (not magnetic) north. The effect of errors in the alignment – and there must always be errors of some sort – is to make a star or planet drift northwards or southwards in the field of view. For high-power or photographic work, where we may want the telescope to guide accurately to within a few minutes or even seconds of arc during the work, the polar axis must be capable of alignment to within a fraction of a degree of the celestial pole.

The declination axis

At right angles to the polar axis is the *declination axis*, on which the telescope tube is mounted. This permits the instrument to be moved in a north-south direction in search of the required object. Once the object is acquired, the declination axis is clamped and should, in theory, require no further adjustment; in practice the image will always drift to some extent, and a slow-motion facility is needed. However, the declination axis plays a secondary role in the success of an equatorial telescope.

Types of equatorial mounting

The bare bones of the polar and declination axes can be assembled in a number of different ways, which means that there are several different types of equatorial mounting for consideration. Only the commonest will be outlined here. They are illustrated in Fig. 24.

The German mounting. This is commercially by far the most common, since it is suitable for both reflecting and refracting telescopes. The declination and polar axes form a 'T', with the declination axis overhanging the north (higher) bearing. The mounting has the advantages of relative compactness and easy accessibility of the whole sky. Its major disadvantage is that the telescope tube tends to foul the pillar in certain positions, and the two axes must then be 'reversed' through 180° in order to bring the instrument round to the opposite side of the mounting. The telescope has to be balanced by a counterweight on the opposite end of the declination axis.

The fork mounting. This is the most compact mounting of all, the tube swinging inside a fork that is fixed to the north end of the polar axis. However, unless the tube is short enough to swing through the fork, the regions of sky near the celestial pole are inaccessible. However, this is a minor disadvantage in most fields of observation, and the fork mounting is probably the most suitable of all for a reflecting telescope. It cannot be used for a refractor because the long tube would foul the pier or mounting in many positions.

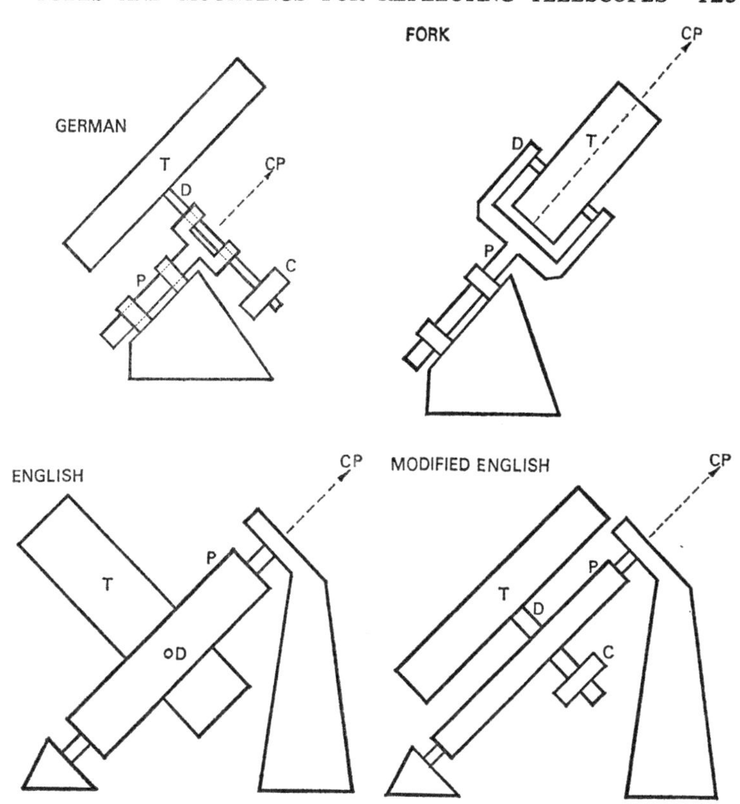

Fig. 24 Types of equatorial mounting. C: counterweight. D: declination axis. P: polar axis. T: telescope. CP: celestial pole

The English and modified English mountings. The German and fork mountings have the undoubted weakness that the telescope's load comes outside the bearings. The English (or polar frame) mounting has a hollow polar axis with the tube swinging inside it between the bearings. In the modified form, the polar axis is solid once more, with the tube mounted to one side and a counterweight on the other. These mountings have been used for many very large reflectors. Their great drawback is that they take up a great deal of space and are expensive in material.

Weight distribution

Balanced or 'inboard' weight counts for a great deal in the rigidity of any mounting, and the most important principle concerning the construction of the essentially unbalanced equatorial mountings – the German and the fork type – is to put weight where it is needed for steadying purposes, and to remove it from where it is undesirable. Thus, in both cases, a massive polar shaft, and the largest possible north bearing – which takes the weight of the whole instrument – provide the best foundation for the telescope, which should itself be as light as possible since it is in an overhanging state. In the case of a German mounting, the declination axis should be as close as is practicable to the north bearing. The overhang in the fork type is considerably more, and therefore requires a more massive construction than the German.

Tube length

The most influential factor in deciding whether or not a mounting will be satisfactory has to do with the length of the telescope tube. Wind, or the observer's interference when focusing or adjusting the slow motions, can set the tube vibrating, and until the oscillations have died out the image cannot be observed properly. The degree of vibration that a tube will suffer depends on several things, but overall length is the biggest factor, With a given mounting, halving the tube length will make the telescope at least four times more 'rigid'. On top of this, the vibrations in a light tube will damp down more quickly than those in a heavy one. When designing a larger instrument, much serious thought should be given to ways of reducing the tube length. For example, a 15cm Maksutov telescope (see page 180), with a tube only about 45cm long but having an effective focal ratio of 15, can be held more steadily on a light tripod and bearing units weighing about a kilogram each than could the equivalent 15cm f/15 refractor, with a tube some 225cm long, set up on a concrete pillar with a total mounting weight too heavy for one person to lift at all. Given equivalent optical performance, there is no doubt which instrument will prove the more satisfactory in use.

Motor drives

The manner in which an automatic drive is fed to the polar axis is another fundamental issue. Having gone to the trouble of making a sound equatorial mount, automatic following transforms the usefulness of the telescope at the cost of little extra work. It is not necessary to have a full 360° drive capability. The scheme normally adopted is to have a worm-and-wheel drive, with the wheel fitted to the polar axis, but this will be smooth and rigid only if the wheel is large enough. The larger the wheel, the better; at all events, its diameter should not be less than about $\frac{1}{8}$th of the tube length. Purpose-built units are very expensive, and a much cheaper but surprisingly effective solution is to fit a tangent drive. An arm projecting from the polar axis carries a hinged nut, through which a length of screwed rod passes. If this rod is turned at the appropriate speed by a geared-down electric motor, an approximately uniform rotation will be imparted to the polar axis as long as the angle between the arm and the rod is close to 90°. As the angle reduces, so the effective speed of the polar axis slows down. Even this effect can be mitigated, to some extent, by using a synchronous electric motor coupled to a variable-frequency unit. These latter are available at reasonable prices (see Appendix A), and make it possible, at the turn of a knob, to speed up or slow down the motor without changing its power. In this way, the observer can compensate for errors in his drive as they occur. Eventually, of course, the arm must be wound back, but the system will be found adequate for most purposes.

Seeking advice

The person wishing to make himself a more elaborate mounting would do well to study as many articles as possible by telescope-makers, and, even better, to visit other amateurs and see their efforts. It is far better to let the other chap make the mistakes, and to profit from his errors! As an introduction to the local amateurs, find out the address of the nearest astronomical society, join, go to their meetings, and seek advice. The writer has yet to meet the keen amateur who is not delighted to show others his equipment!

CHAPTER EIGHT

Optical Alignment and Accessories

The production of a first-class objective must form the first aim of the telescope maker, and a firm and manoeuvrable stand is equally important. There are, however, other essential links in the chain. If the diagonal mirror and eyepiece are not up to standard, the performance of the main mirror will suffer; and there is still the matter of accurate optical alignment to attend to.

The diagonal

It is often said that an optically flat mirror is harder to produce than any other surface shape, but this is not strictly true. A flat surface can be considered as forming part of the surface of a sphere of infinite radius; if we were asked to produce a 15cm f/10 sphere with a tolerance of $\pm$0.15mm on the radius of curvature, it would be as difficult as producing a flat of the same diameter accurate to $\pm\frac{1}{10}$th of a wavelength of light. Since, in practice, we do not mind if the mean radius of curvature of our mirror wanders by several millimetres during the figuring process as we play off one zone against another, the job is easier to do. Generally speaking, the amateur would be advised against trying to make his own diagonal, certainly until he has gained experience.

The beginner will therefore buy his diagonal from a reputable firm making astronomical optics. Do not place much faith in the 'flat' mirrors taken from miscellaneous optical equipment unless their quality is guaranteed against a refund, since they were probably not made to astronomical tolerances. Some concerns offer unworked plate glass, either elliptical or rectangular, as optical flats! Plate glass is polished with felt pads and has an optical 'ripple' to it, besides rarely being generally flat enough for the purpose.

The standard size of minor axis (the narrow measurement across the face) of a flat for a 15cm reflector is 3.5cm, which is large enough to transmit light from the mirror to all parts of a normal low-power field. The closer the eyepiece is to the side of the tube, the smaller the flat need be, while the shorter the focal ratio of the mirror, the steeper the cone of light and the larger the flat needs to be to give full illumination across the same linear field (as measured with a ruler) at the focus. The size of the minor axis m of a diagonal situated a distance a inside the focal point of a mirror diameter D and focal length F, to give a fully-illuminated field of linear diameter B, is given by

$$m = \frac{Da + B(F - a)}{F}$$

The surface accuracy of a flat becomes of critical importance only when high powers are being used. Under such circumstances, the observer is examining an object right on the axis of the mirror, so that the field of view is effectively zero. This means that only the central part of the flat, whose effective minor axis is equal to Aa/F, is being used. In the current case, if we suppose the focus to be 10cm outside the tube, we have a=21cm (say), and m becomes 15 x 21/150=2.1cm. Thus, less than $\frac{1}{3}$ of the area is being used for critical work. This is why most opticians offer '$\frac{1}{4}$-wave' accuracy for their diagonals, since the effective accuracy, when it matters most, is three times better than this because of the reduced working area.

Aligning the optics

Let us now consider the problem of aligning our two mirrors. The initial work is best done in daylight. It will be assumed that the axis of the flat mount has been placed centrally in the tube by correct fixing of the arm or vanes. The next task is to bring the diagonal on to the axis of the eyepiece tube,

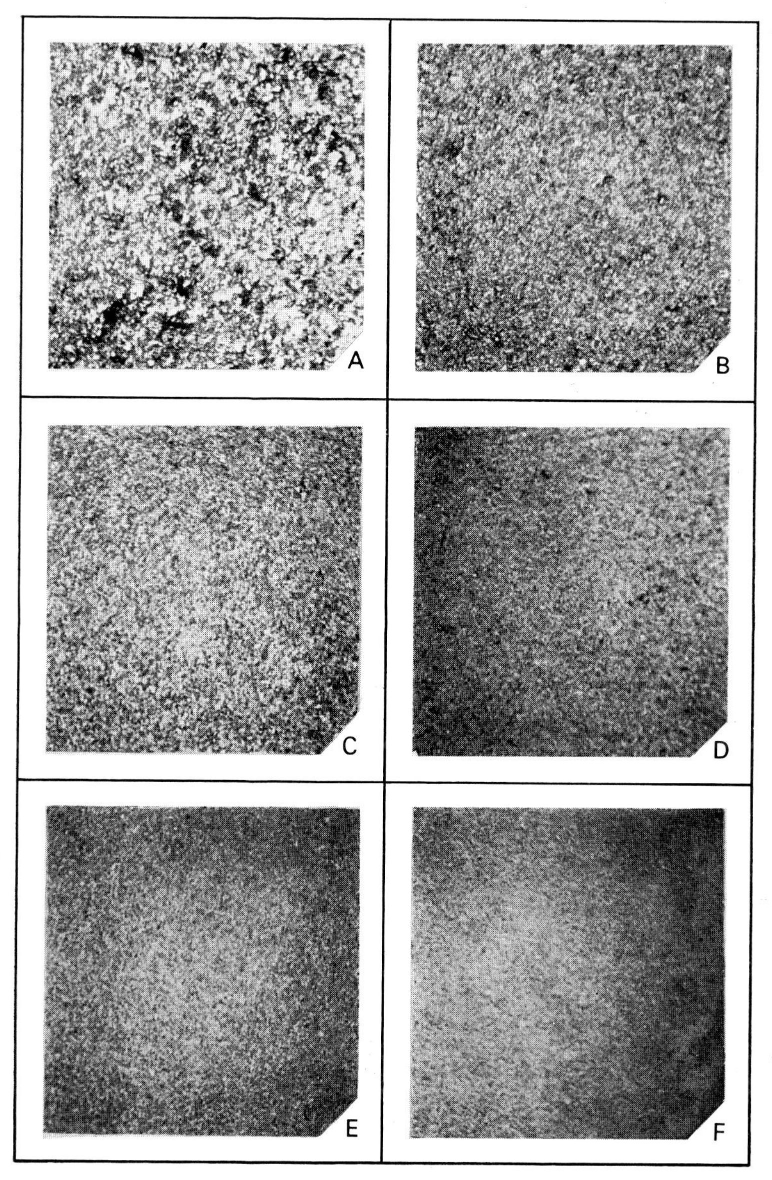

1. Pits left by different grades of abrasive. A-D: carborundum grades 80, 150, 280, 320. E & F: aloxite grades 225 & 125

2. A pitch polisher

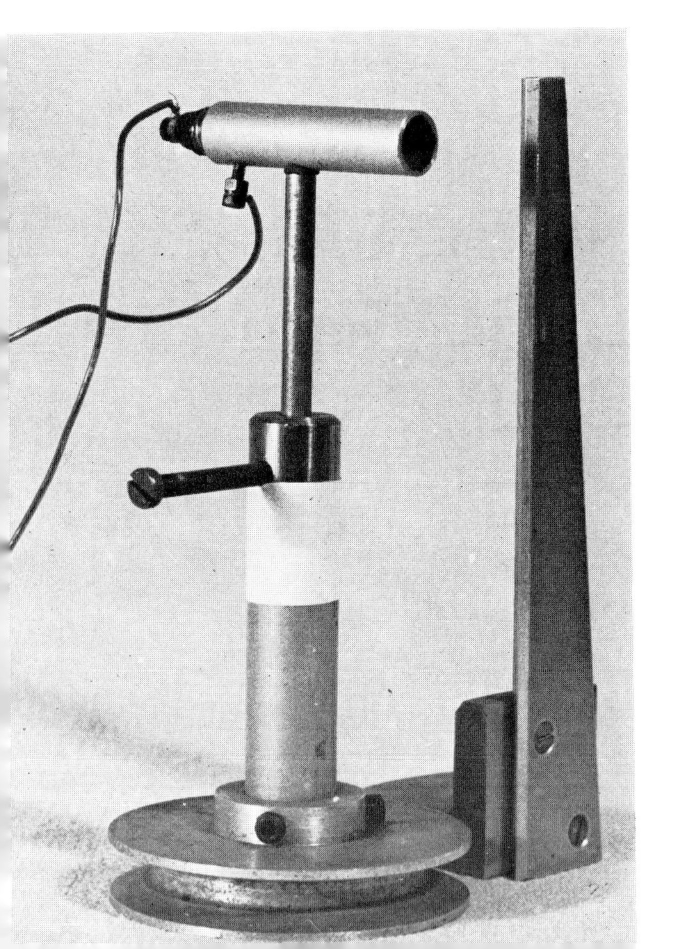

3. A Foucault tester

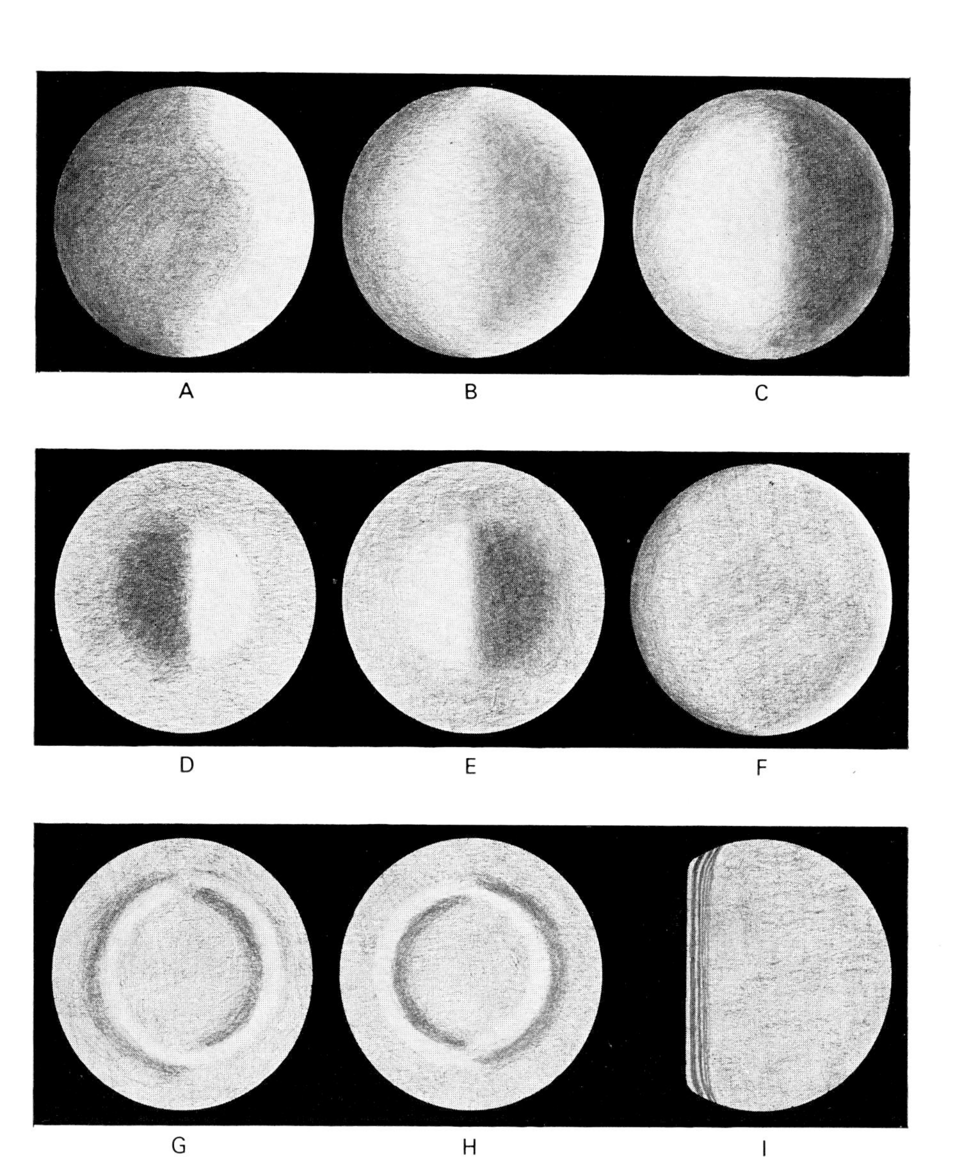

4. Knife-edge shadows in the Foucault test. A-C represent the appearance of a prolate mirror with the knife-edge at the centre of curvature of the inner, 70 per cent, and marginal zones respectively. D & E show a spherical mirror with central hill and central hollow. F shows a generally spherical mirror with a rolled-back edge. G & H represent raised and depressed zones respectively on a spherical mirror. I shows the diffraction lines to be seen in front of the knife-edge when some distance inside the centre of curvature; a turned-down edge of about $\frac{1}{4}$-wave is shown.

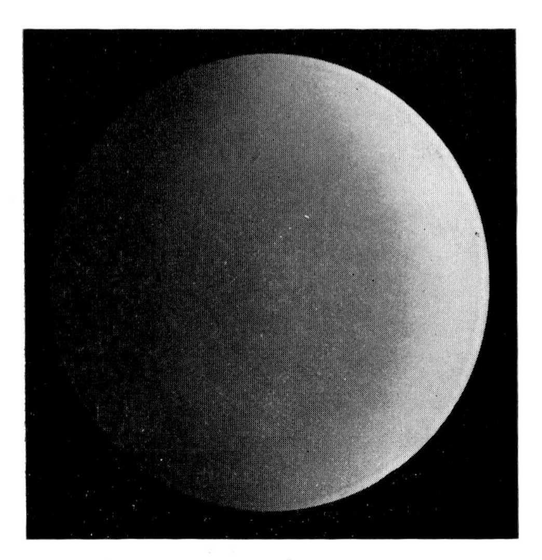

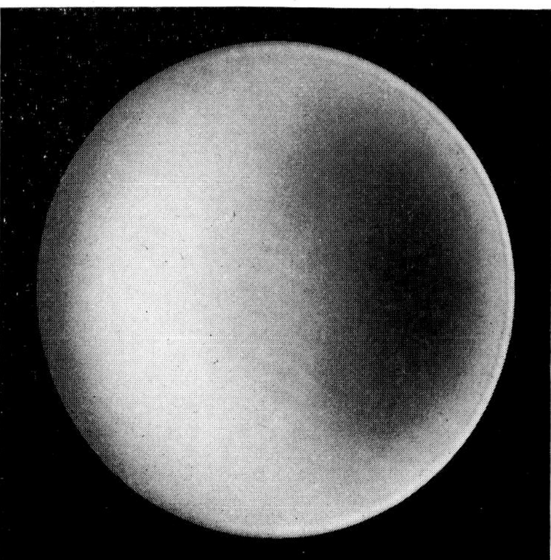

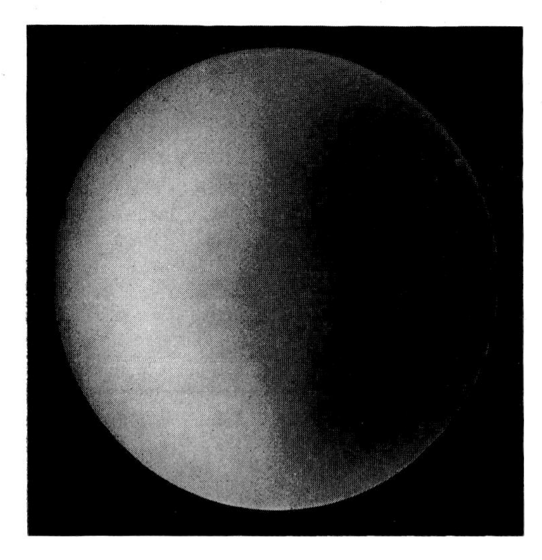

5. Foucault test of a prolate mirror. Three test photographs of an early 15cm f/4 paraboloid made by the writer, taken at the centre of curvature of the inner, intermediate and marginal zones. A slight central hill, with a faint depressed zone around it, is well seen in the first photograph. This is characteristic of a mirror figured, as this one was, using the W-stroke

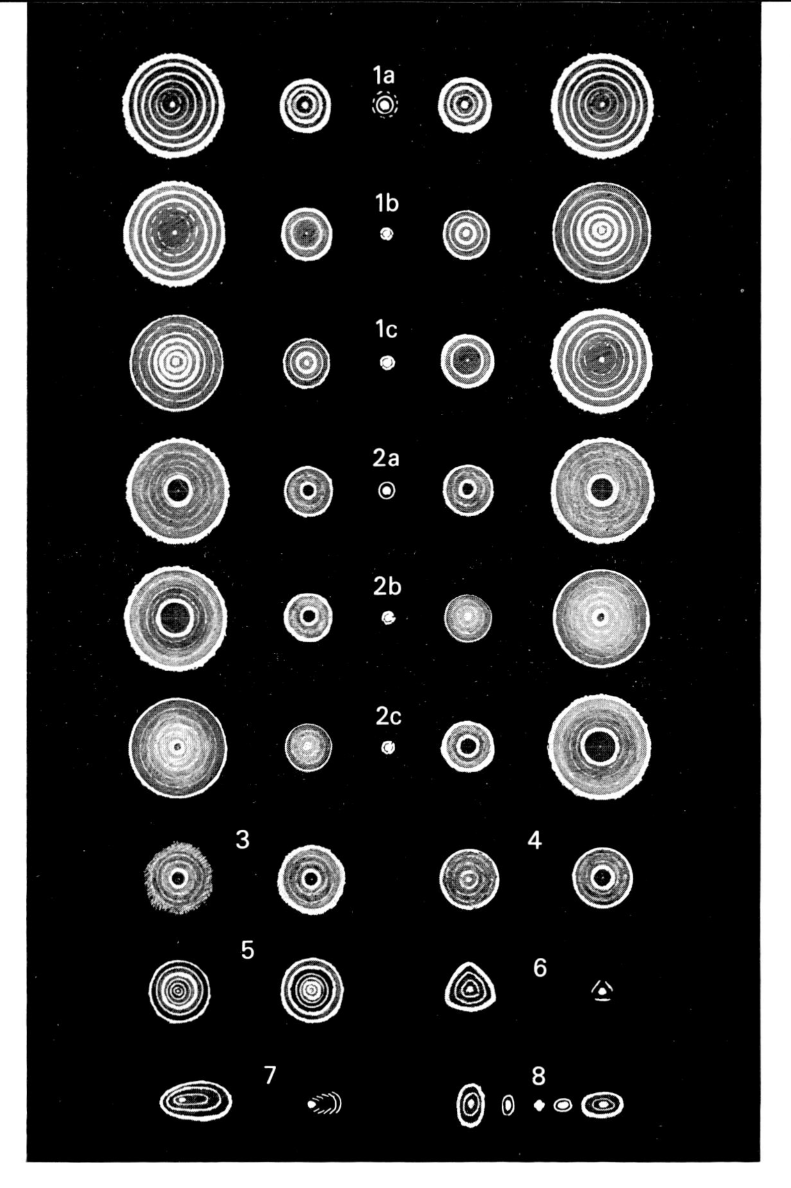

6. Star images in telescopes of different quality. 1 and 2 show the images to be seen in a refractor and a reflector respectively as a high-power eyepiece is moved from inside the focus through to an equal position outside the focus. a=perfect correction; b=under-correction; c=over-correction. Diffraction rings are relatively weak in the defocused image of a reflector unless it is of long focal ratio. 3: Generally well-corrected mirror with turned edge, inside and outside focus. 4: Generally well-corrected mirror with a central hole, inside and outside focus. 5: Refractor with zonal error in the 50 per cent region. 6: Typical triangular appearance of defocused and focused image, due to incorrect three-point support of the objective. 7: Seriously comatic image, defocused and focused. 8: Typical appearance of astigmatism as the eyepiece is taken through the focus

7. Testing for wedge in a lens

8. Trepanning a Cassegrain mirror. A simple rig made by the writer and powered by a 1/6th horsepower motor. The vertical shaft is free to descend as the hole is cut. In this photograph, a 50mm hole is being cut in a 30cm disc. Some spare trepans can be seen to the right

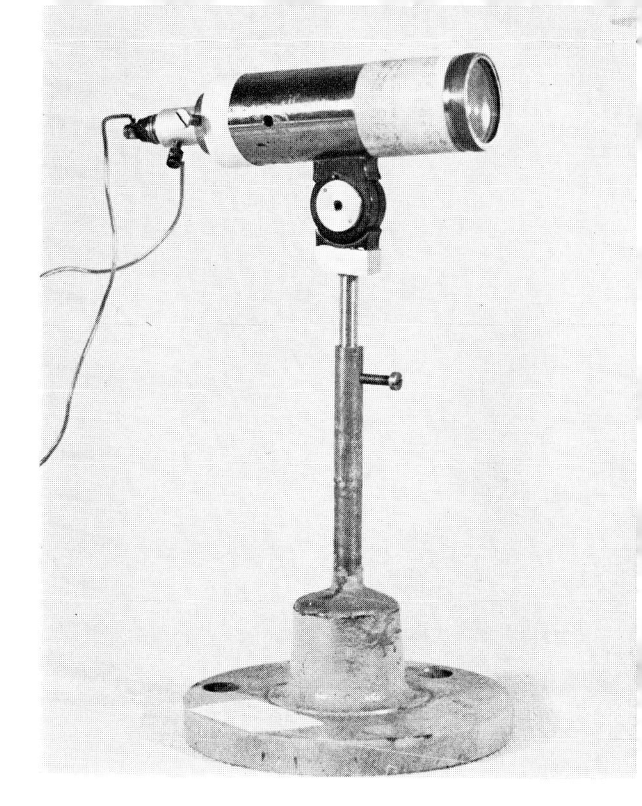

9. A Dall null tester, made by the writer. The knife-edge is not shown

10. An accurate sphereometer

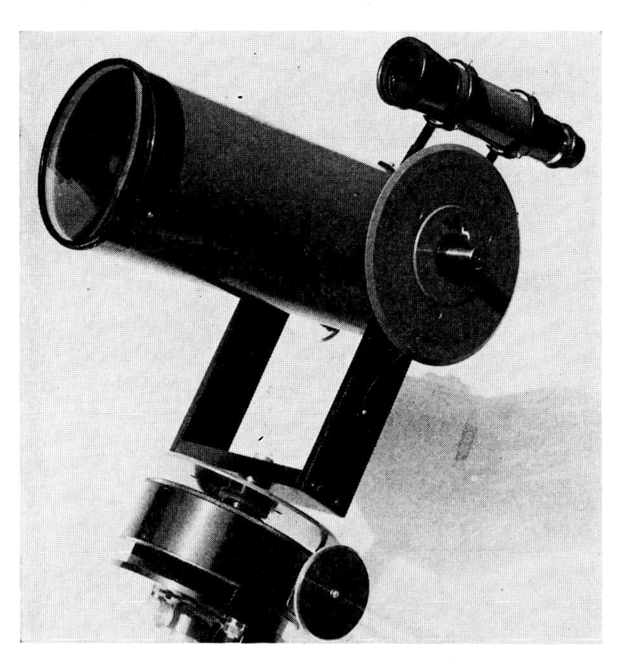

11. *Above:* A Maksutov shell with its tool. The polisher was poured on a similar casting

12. A Maksutov telescope. This is a well-equipped commercially-made instrument of 15cm aperture

which should be, within very narrow limits, at right angles to the side of the tube. This can be done by eye-estimation if a cardboard disc, with an accurately-placed central hole to allow the eye to be positioned axially, is fitted into the eyepiece end of the focusing mount; the circular outline of the flat should be concentric with the inner outline of the eyepiece tube (Fig. 25). If it is not, lengthen or shorten the

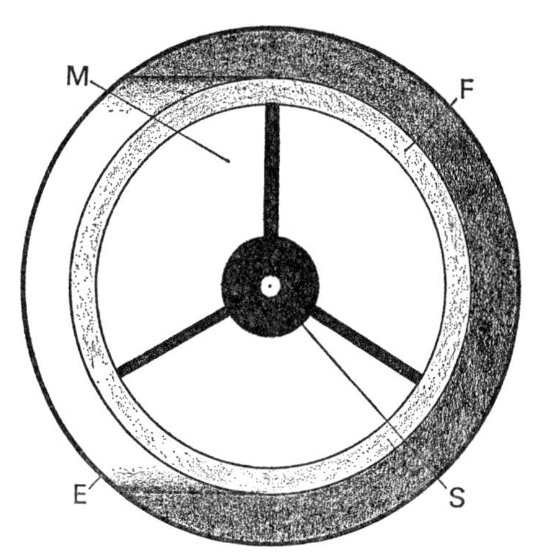

Fig. 25 Aligning a Newtonian telescope. This represents the view through the eyepiece tube, with the eyepiece removed. E: outline of eyepiece tube. F: surface of flat mirror. M: image of main mirror in flat. S: outline of flat and supports reflected in the main mirror

adjusting screws at the base of the flat mount. Using the same screws, bring the reflection of the primary mirror central within the outline of the flat. This part of the alignment operation is now complete. To avoid the confusing reflections from the primary, it may be easier to adjust the flat with the primary covered.

The primary mirror is now adjusted with the nuts holding its cell to the base of the tube. Many makers simply have compressed springs between the back of the cell and the base, but the writer prefers the positive locking of the suggested arrangement. The inner nuts can be reached through the central hole in the base, and final tightening can be done with a spanner, although knurled nuts for hand-tightening would be preferable. The mirror is in alignment when the circular black outline of the flat is central in the mirror's own outline seen in the flat. If the flat's silhouette appears to one side in the reflection of the main mirror, the side of the mirror on the opposite side should be advanced. A friend to observe, or to adjust, will be a great help.

Final adjustment is performed at night, on a star, using a high-power eyepiece. If the mirror's optical axis is not passing through the centre of the eyepiece, the centralised image will show a slight flare to one side: the indication of coma. If the maladjustment is only slight, it may be imperceptible in the focused image, but the defocused discs will still show eccentricity. In a parabolic mirror, the point of the flared image indicates the direction of the true axis, so the mirror is adjusted to move the image tail-first away from the centre of the field. The star is centralised once more, using the slow motions, and the process repeated until the axial coma is eliminated. (See Plate 6.)

Distortion

If the image refuses to come symmetrical even when the best alignment is secured, there must be some distortion of one or both of the optical surfaces. It may be that the primary mirror is being 'pinched', but using glass of the recommended thickness it would require considerable pressure to achieve this. To check, aim the tube almost horizontal and set up the illuminated pinhole at the centre of curvature, examining the image with a high-power eyepiece. If the extra- and intra-focal discs are perfectly round, and without flare, the fault cannot lie here, and the flat should be suspected. A return to the maker for checking is the best course, once it has been established that the diagonal mounting is not subjecting the glass to any undue pressure.

A persistent astigmatic appearance of the image, in which the defocused discs are elliptical rather than circular, with the major axes at right angles to each other, indicates a serious optical fault or gross misalignment. In seeking the culprit, do not forget that most human eyes are, to some extent, astigmatic. If the eye is to blame, the use of higher powers will minimise the astigmatic appearance since the narrower eyebeam is less affected by a defective cornea; if the instrument itself has an astigmatic component, increased magnification will make the fault appear more obvious.

Magnifying powers

The final link in the telescopic chain is the eyepiece, whose purpose is to magnify the primary image by allowing the observer to view it from a much closer distance than would otherwise be possible. To take an example, the lunar image formed by a 15cm f/10 telescope is about 13mm across. If we view the image of a distant object formed by an objective from a distance equal to the focal length of the objective, it appears the same angular size as the real object. Therefore, if a short-sighted person, who can perhaps focus his eyes down to a distance of 10cm, views the 13mm-diameter image of the moon from this distance, he will see a moon 150/10= 15 times larger than the one in the sky. An eyepiece, when placed a distance equal to its focal length away from the image being examined, emits a parallel beam into the observer's eye; he can therefore view the image from very close range while keeping his eye comfortably relaxed at the infinity focus.

We have already seen (page 34), that the *lowest* useful magnifying power of any telescope is obtained from 1.25D (cm) or 3.3D (in.), where D is the aperture. With moderate and large apertures, the limit on *high* magnifications is set principally by atmospheric turbulence – supposing, of course, that the object being observed requires a high magnification. It is the rule, rather than the exception, for the air through which we must view astronomical objects to consist of layers at different temperatures, with irregular mixing occurring at their boundaries. The refractive index of air varies with temperature – the shimmer over a heated road in summer is an example of this – and under such conditions of poor 'seeing' the sharp outlines of the image will be lost. The magnification required on such an occasion is only that needed to show well the detail that the atmosphere allows to be seen; and it will be found, because of the loss of contrast associated with more powerful eyepieces, that less rather than more detail will be seen if the magnification is increased beyond this practical limit. On the relatively rare occasions of almost perfectly steady atmosphere, planetary and stellar detail will continue to benefit from increased magnification until either the image becomes too faint or we reach the limit of detail imposed by diffraction. The latter limit is commonly taken to be about 40D (cm), but few astronomical telescopes can ever benefit from magnifications exceeding about 20–25D because of the slight atmospheric wobbling that occurs even on the best nights. Hence, an upper limit of ×350 or so would be applicable to our 15cm reflector, corresponding to an eyepiece focal length of about 4mm. The whole subject of seeing and magnification is gone into more thoroughly in the writer's *Astronomy for Amateurs* (Cassell, 1969, page 57 *et seq.*).

The merits of any particular eyepiece are to be judged from the following aspects.

Spherical and chromatic aberration

The effect of these aberrations varies inversely as the focal ratio of the telescope with which the eyepiece is used. Generally speaking, chromatic effects are the more serious. At f/10, most eyepieces are satisfactory in both respects, but at f/5 and below the choice is restricted to the highly-corrected types if the mirror's quality is not to suffer.

Apparent field of view

This is the angular diameter of the bright circle of light seen when the eyepiece (or the complete instrument) is pointed to the day sky with the eye in the normal viewing position. In some eyepieces it is as small as 30°; elaborate types have achieved nearly 90°. A field of 40–50° is typical. The real field of view of the telescope is given by the apparent field of the eyepiece divided by the magnification. More important than the diameter of the apparent field of view is the quality of the definition near the margin; many ultra-wide field eyepieces are so afflicted with off-axis astigmatism that stars near the edge of the field appear as lines rather than points. Most eyepieces suffer from field curvature, which means that their focal plane is not flat; in some cases this can be put to advantage, if the curvature of the field of the objective approximately coincides with that of the eyepiece.

Eye relief

The eyepiece forms, somewhere in the eyebeam (see page 34) and of the same diameter, an image of the objective known as the *Ramsden disc* or *exit pupil*. If the Ramsden disc does not fall in or near the observer's pupil, he will not be able to see, simultaneously, the full field of view afforded by the eyepiece. The distance from the outer face of the eyepiece's eye lens to the Ramsden disc, which is known as the *eye relief*, varies with the type of eyepiece and the focal length. High-power eyepieces with short eye relief are uncomfortable to use, since the eye must be brought almost into contact with the eye lens.

Ghosts

Some light is wasted in any eyepiece or lens system because of the weak reflection that occurs at any air-glass surface. Glass normally reflects about 4 per cent of the incident light, and in an eyepiece containing a number of separate elements the light-loss in this way can become as much as 25 per cent. 'Blooming' the glass surfaces with a very thin coating of magnesium fluoride, or some other compound with a refractive tive index intermediate between that of glass and air, considerably reduces this unwanted reflection, although it can never be eliminated altogether. If any of these reflections come to a focus near the focal plane of the eyepiece, a bright object in the field of view will have a 'ghost' companion. Such ghosts are particularly confusing when a faint star is being sought in the vicinity of a bright one, and they can also cause loss of all-important contrast in planetary observation. For these fields of work, ghost-free eyepieces are to be preferred.

Eyepiece designs

A single lens of short focus will constitute an eyepiece of sorts, but the field of view of good definition will be small, and it will not be achromatic. All eyepieces worthy of the name therefore consist of at least two lenses or elements. If these elements are separated, the one nearest the eye is called the eye lens, and the one facing the objective is called the field lens because part of its task is to widen the field of view that would be obtained were the eye lens alone used. Both 'lenses' can consist of single elements or be achromatic combinations. The terms 'negative' and 'positive' are used to distinguish between those eyepieces whose focal plane lies between the lenses (negative), and those whose focal plane lies somewhere in front of the field lens (positive). A circular aperture, known as the field stop, is placed at this position to give a sharply-defined margin to the field of view. Eyepieces with the focal plane very near the front surface of the field lens are objectionable because of the sharp relief into which specks of dust are thrown!

Two-element eyepieces

The *Huyghenian* and *Ramsden* eyepieces (Fig. 26) both consist of plano-convex lenses; in the Ramsden type they are identical, with their flat faces pointing outwards. In practice there is little to choose between them, since they are relatively poorly corrected for spherical and chromatic aberration, and are used mainly with telescopes of large focal ratio (f/10 and above).

The *achromatic Ramsden* eyepiece is often advertised. Both single elements are replaced by cemented achromatic com-

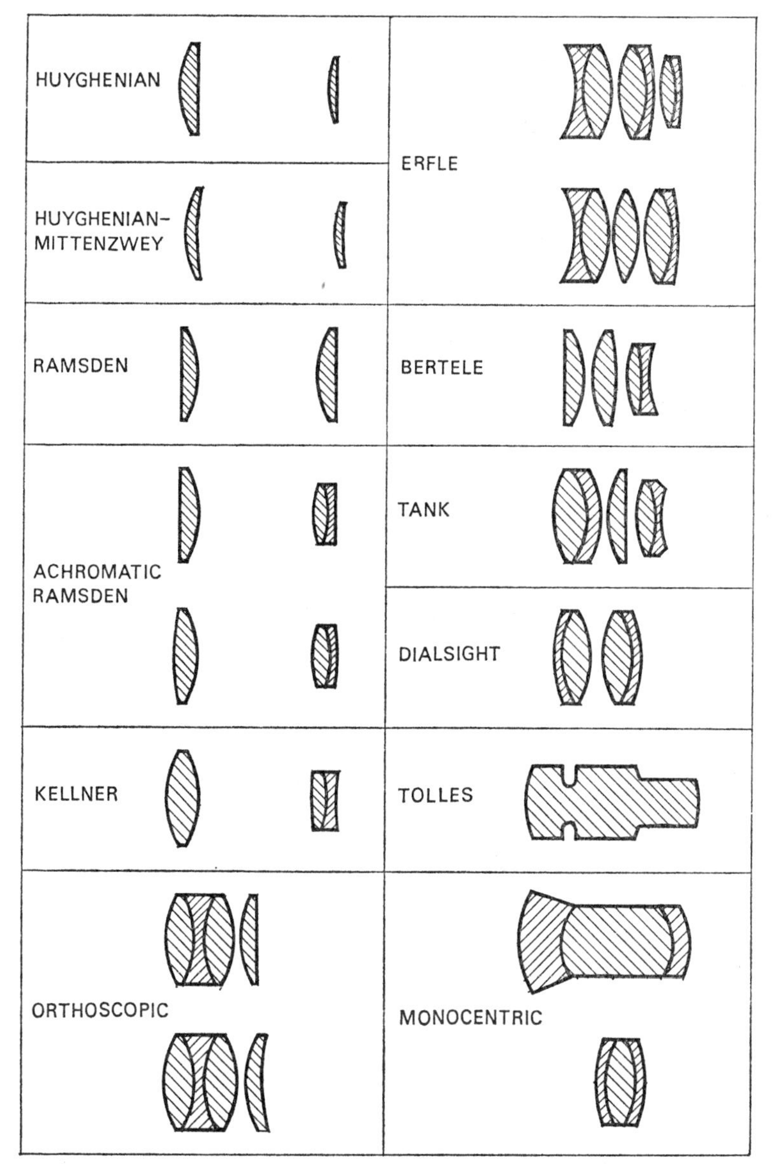

Fig. 26 Types of eyepiece. In all cases the field lens is on the left.

binations, giving much improved colour correction and spherical aberration. These eyepieces can be used with telescopes as fast as f/4, if they are well made. The *Kellner*, effectively a Ramsden eyepiece with an achromatised eye lens, is rarely seen today, although for some reason many achromatic Ramsdens parade under its name.

The most common type of eyepiece offered for use with reflecting telescopes is the *orthoscopic*. In this family of eyepieces the distinguishing feature is a single eye lens and a double or even triple field lens. Orthoscopics enjoy the manifold advantages of wide apparent field, excellent correction, and long eye relief, the latter being particularly important when high magnifications are being used. It is the most universal type of eyepiece, giving acceptable results at all ranges of the magnification scale. Unfortunately it is relatively ghosty, although some types are better then others. Allied to the orthoscopic is the *Erfle*, a more modern design containing three achromatic elements and capable of giving apparent fields of 80° or more; these eyepieces are popular for low-power work.

The *Dialsight* or *Plossl* eyepiece, also popular for low-power use, has two symmetrical achromatic elements, and gives exceptionally long eye relief, making it a useful eyepiece for spectacle wearers. Its freedom from ghosts enhances its value as, for example, a comet-sweeping eyepiece. The writer does not know of any commercial eyepieces of this design, but it is readily available on the surplus market in fairly long focal lengths.

Cemented or 'solid' eyepieces

These are the choice of the discerning planetary or double-star observer who is working at high magnification with a Newtonian telescope. There are only two air-glass surfaces, and hence very high light transmission, and they are completely free from internal reflections that can reduce the contrast of faint markings enough to make them invisible. They all have a relatively small field of view (about 30°), but the diameter of the field of critical definition in a wide-field eyepiece may be no more than this. The *Tolles* type is made from a short cylinder of glass, with convex ends; it is relatively cheap to make, but suffers from very poor eye relief, so that it is difficult to see all of the field of view at once. Superior, and extremely expensive, is the *monocentric*: a cemented triplet with satisfactory eye relief and excellent correction. A monocentric of good quality, with a focal length of 6–9mm, is probably the finest high-power eyepiece available – if one can be found! Its outer lenses being of soft flint glass, it should be looked after with the most scrupulous care.

Selecting eyepieces

The above remarks apply only to well-made eyepieces of the various types. It is often assumed that *any* eyepiece is representative of its pattern, but this is not so, because their performance can be affected by faults in design or execution. The orthoscopic type, in particular, embraces so wide a range that large variations in performance must be expected. The accompanying table summarises the more important characteristics of the eyepieces just described.

EYEPIECE TYPES

Eyepiece	*Aberrations*	*Field*	*Eye relief**	*Ghosts*	*Remarks*
Huyghenian	Poor	40°	0.3	Fair	Very suitable for refractors of normal focal ratio
Ramsden	Fair	35°	0.25	Poor	Similar to Huyghenian
Achromatic Ramsden	Good	40°	0.25	Fair	Usable on normal Newtonians
Kellner	Good	45°	0.45	Poor	Rarely found; superseded by orthoscopic type
Orthoscopic	V. good	50°	0.8	Poor	Usable on any telescope: the 'universal' eyepiece

*Given in terms of the focal length of the eyepiece.

Eyepiece	*Aber-rations*	*Field*	*Eye relief**	*Ghosts*	*Remarks*
Erfle	V. good	70°	0.5	Poor	Popular for low-power work
Dialsight (Plossl)	Good	40°	0.8	Good	An excellent low-power eyepiece
Tolles	Good	40°	0.0	None	Usable down to f/7. Fine for planets, but inconvenient
Monocentric	Excellent	30°	0.8	None	The best high-power eyepiece

Due to the tremendous influx of Japanese eyepieces that started a decade or more ago, British-made oculars have become almost a rarity. This is a pity as regards the low magnification range, because the imported eyepieces all have a 24.5mm push-in fitting, which means that the maximum diameter of the field stop is about 20mm, considerably less than the maximum obtainable with the British 1¼-inch R.A.S. threaded type. Far better results will be obtained using a large-diameter eyepiece of the orthoscopic or Dialsight form taken from surplus equipment, with a home-made adaptor to fit it to the focusing mount.

It will be found, in general, that three eyepieces will give satisfactory service. More will certainly be acquired eventually, but it will probably be found that most of the eyepieces in the collection are rarely used. Magnifications of about ×40 (40mm focus), ×125 (12mm) and ×250 (6mm) will cover most needs.

It may be asked why we do not follow the earlier conclusion and aim at the low power of about ×19 permitted by the limiting size of the Ramsden disc. The reason is that no normal diagonal size or eyepiece fitting could transmit light across such a large field; with an eyepiece of 45° apparent field the real field of view would be 2½° across, representing a linear diameter of about 7cm. The flat would therefore have a minor axis over half the diameter of the primary! On top of this consideration, the off-axis aberrations of the mirror, and its curvature of field, would ruin the definition at the field margins. A field of the order of 1°, at about ×40, is the practical limit of the ordinary 15cm Newtonian; if very low powers are required, it would be necessary to have a short focal ratio to reduce the focal image scale, and to adopt a different design of telescope with better off-axis characteristics.

The Barlow lens

New eyepieces are no longer cheap, and the widely-available Japanese orthoscopic type, which are made in a useful range of focal lengths, can currently cost £15 or more. A way of making one eyepiece do the work of two is to employ a Barlow lens. This is a diverging achromatic lens which, when placed a small distance inside the focus of the telescope, effectively increases the focal length while moving the focal point only slightly. Most Barlow lenses have a negative focal length of about 10cm, and are designed to give an amplification of 2, so that an f/10 instrument becomes effectively f/20; the magnification of each eyepiece is therefore doubled. In this instance, the Barlow would have to be positioned 5cm inside the focus, which would be re-formed 5cm beyond its original position (Fig. 27). 'Zoom' or variable-power eyepiece units incorporate a sliding Barlow lens giving a range of powers. A Barlow lens is an attractive addition to the short-funded observer's kit, but most astronomers who habitually use the high powers that it allows would ultimately prefer to buy the proper eyepiece for the job.

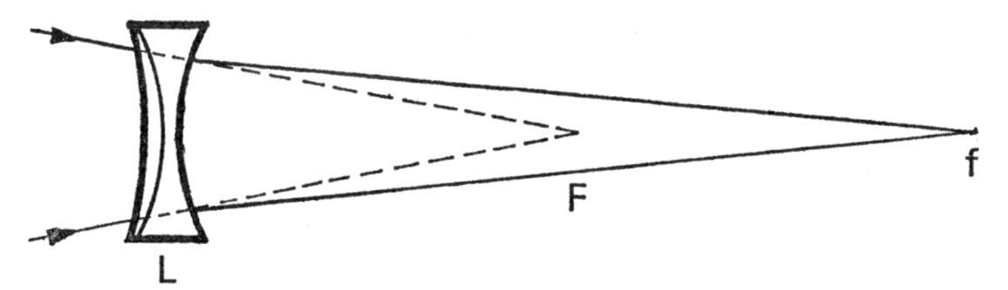

Fig. 27 The Barlow lens. F: original focal point of telescope. f: new focal point produced by the Barlow lens. The amplification of the lens is given by Lf/LF

CHAPTER NINE

Designing and Making an Object Glass

Amateur telescope makers will encounter a formidable body of opinion advising them against trying to make an object glass. They will hear that it is a long and tedious job, that the glass is expensive, and that professional facilities are needed for working the glass and testing the finished lens. This is no longer the case, if, indeed, it ever was. Optical glass is readily available, and the supplier can cut the blanks circular to whatever diameter is required. The 'special tools' needed amount to a couple of plate glass discs of the same diameter as the lenses. Two small testing rigs are the only special requirement, and they could both be made in one or two evenings, with no special facilities.

Why, then, the mystery? Perhaps it dates back to the days when optical glass was of unpredictable quality and required special skills in testing and figuring if the results were to be worthwhile. Possibly, too, telescope makers in those days were reluctant for their object glass designs to be widely published and copied. These days there are designs and optical glasses in plenty, and the purpose of this chapter is to show that anyone who has made himself a mirror that he is pleased with can make himself an object glass – and probably one giving better images. It is about time someone spoke up for the refractor: as far as the writer is aware, nothing serious has been published in this country on amateur object-glass manufacture for the past half-century!

Lenses versus mirrors

Certainly the work is longer, for there are four surfaces to grind and polish. But this is merely a matter of persistence, and some time can be saved by choosing a design in which one of the convex curves of the crown matches the concave curve of the flint, so gaining the benefit of two smoothed surfaces from one operation. And the polished but unfigured combination will already be giving excellent results if a focal ratio of about f/15, which is normal for small refractors, is chosen. The reason is that, as we have seen, errors of figure in a lens are only one quarter as important as those on a mirror. It is perfectly possible to figure an object glass so precisely that no residual error whatever can be detected. This can rarely be said of a mirror, even a professionally-made one. A lens, moreover, is insensitive to thermal effects and in any case loses its heat quickly, and of course it requires no reflective coating. Scattering of light from dusty surfaces, which can reach monstrous proportions in an ill-tended Newtonian, is reduced by the fact that no diagonal is required; furthermore, since dust tends to scatter light back *into* the oncoming beam, it has a less obtrusive effect in refracting than in reflecting systems. Therefore a refracting telescope will normally give a cleaner image than a reflector, and very likely defines better. An extra advantage of the refractor is that no diagonal and supporting system is interposed in the light path to diffract light out of the Airy disc and into the rings, which lowers contrast in the same way as does a slightly imperfect figure. A diagonal of only $\frac{1}{4}$ of the aperture of the primary, as recommended, does not degrade the image seriously, but it represents an extra advantage for the refractor.

Secondary spectrum

The refractor suffers from the disadvantages of imperfect achromatism and bulkiness. The latter arises from the former, since, to reduce the effect of secondary spectrum, a long focal ratio is needed. The existence of secondary spectrum arises from the fact that an achromat consisting of two components cannot normally be designed to bring more than two separate colours to the same focus. We can represent the situation in

Fig. 28. A single lens produces a series of images, from violet (nearest the lens) to red – other wavelengths are also brought to their respective foci, but are invisible to the eye. By achromatising the lens, the designer arranges for the two selected colours to be brought to a common focus. These colours are defined by their wavelength in nanometres*. The approximate limits of human visibility are 420nm (blue) and 700nm (red), with maximum sensitivity at about 560nm (yellow), and the wavelengths often chosen for coincidence are 486nm and 656nm. When this is done, the yellow light to which the eye is most sensitive comes to a slightly shorter focus than the red-blue point, while the deep red and violet light, to which the eye has little sensitivity, focuses considerably further back. It is the combination of the slightly out of focus red and blue rays, and the violet, which gives the characteristic purple halo of the secondary spectrum.

The seriousness of the secondary spectrum varies, for an object glass of given focal ratio, with the square of the aperture, and we can incorporate this in the rule-of-thumb

$$f/ \nless 1.2D$$

for the fastest permissible focal ratio with any aperture. Thus, a 5cm object glass can be made as fast as f/6, while one of 20cm aperture would be limited to about f/24. Exceeding these limits will mean that the residual chromatic aberration (yellow to red-blue) will be significantly beyond the Rayleigh limit. On the other hand, this restriction ceases to be so important if the objective is to be used only for low powers and wide fields; the writer has made satisfactory 12cm f/5.5 object glasses for wide-field telescopes, and in apertures larger than about 15cm one is more or less forced to disregard the formula in the interests of a manageable tube length. To take an extreme case, the Yerkes 40-inch refractor works at f/20 although the formula indicates f/120! In these cases, it is a

*A nanometre is a millionth of a millimetre or a thousandth of a micron, and is ten times larger than the Angstrom unit, which used to be used for wavelength measurements.

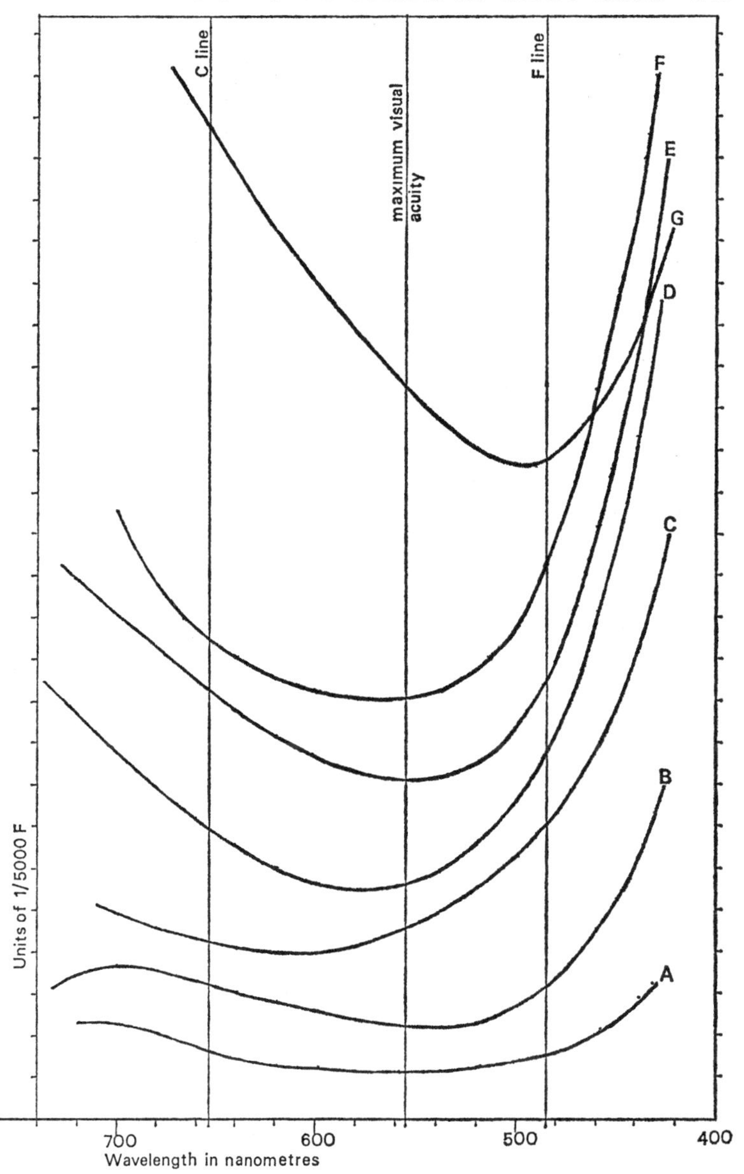

Fig. 28 Achromatisation curves for different object glasses. This diagram shows the variation of focal length in different colours; distance from the object glass increases up the page. The minimum focal length is always aimed to fall near the colour of greatest visual acuity. A: 12cm f/20 Cooke photovisual triplet. B: 8cm f/ 17.5 Dall 2-lens apochromat. C: 8cm f/17.7 Zeiss 2-lens apochromat. D & E: typical 2-lens object glasses by Hilger and Ross respectively. F: the 36-inch (91cm) f/19 Lick Observatory refractor. G: typical correction for a photographic objective. The triplet A represents the closest approach to a perfectly achromatic system, with the 2-lens apochromats (using special glasses and steep curves) offering a middle solution (Courtesy H. E. Dall)

matter of personal taste how much spurious colour is acceptable, and whether the superior defining power of a large refractor is outweighed by the residual tints in which it frames its images. It can be said, quite safely, that the 10cm f/15 object glass advocated in this chapter will be well within all limits, and its definition will be a revelation to anyone used to observing with a mediocre reflector.

Refraction and dispersion

The principle behind the achromatic objective depends on the relationship between *refractive index* (N) and *dispersive power* (V) of the two types of glass used. Everyone interested in optics knows of the power of a glass prism to disperse white light into its component colours; if we take two identical prisms and arrange them as in Fig. 29, the second one recom-

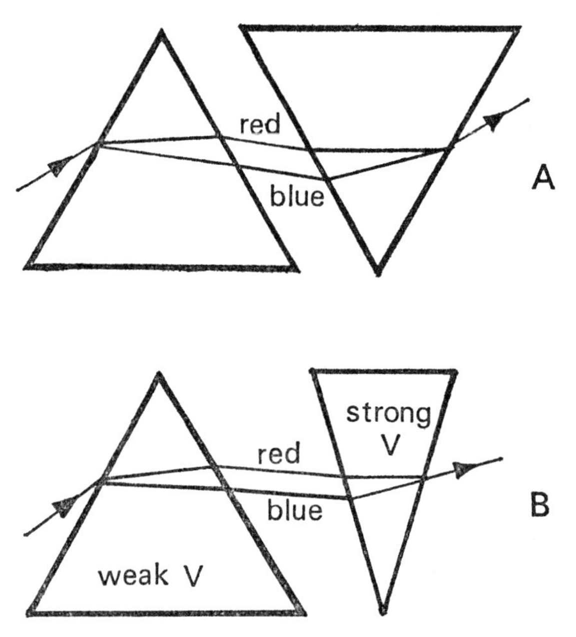

Fig. 29 Dispersion and recombination of light. The prisms in A are identical in all respects. In B, a weakly-dispersive crown prism is matched against a narrower prism of strongly-dispersive flint glass

bines the dispersed light and the original white beam is regained. However, such a system has no *power,* because the dispersive powers of the two glasses are the same. This means that the light passing through the second prism is turned through the same angle (but in the reverse way) as it was when passing through the first. Therefore, the direction of the emergent beam is the same as that of the incident beam; the only resultant effect is that it has been slightly displaced.

Within limits, the amount by which a beam of light is dispersed into colours by a prism depends on the angle of the prism, as well as on the dispersive power of the glass. Therefore, a small-angle prism of strongly dispersive glass may produce as wide a band of colours as a fat prism of weakly-dispersive glass. Suppose, then, that we take two prisms of glass whose dispersive powers are markedly different, and try to repeat the experiment to produce white light from a beam passing through both of them. It will be found (Fig. 29), if this is achieved, that the angles of the prisms do indeed bear some inverse relationship to the dispersive powers of the glasses. On top of this, we find that the emergent beam is no longer travelling in the original direction. The new combination of glasses has the power to refract light through an angle, and so can be used, if formed into a pair of lenses, to form an image.

It will be noted that the refractive index of the glass, or the degree to which it is able to turn light (whatever its colour) through an angle, does not affect the argument. This is because N differs little from one type to another, whereas V varies greatly. It would not matter if N were the same for both prisms, as long as V differs. The two great families of glass, the crowns and the flints, are, generally, differentiated from each other in terms of their dispersive power, although there is a tendency for flint glass to have a slightly higher refractive index than crown. Typical values for the two types would be $N=1.52$, $V=60$ for crown, and $N=1.6$, $V=36$ for flint (note that a lower value for V means greater dispersive power). However, the designer can choose from dozens of different types of glass, with N ranging from 1.45 to 2, and V from 20 to 85.

Designing an achromatic combination

When the glassmaker measures the V value for a sample of glass, he is, in effect, telling the optical designer how far apart two given colours will come to a focus if a simple converging lens were made from the glass in question. Of course, he must state the wavelengths he has based his value on. A very common choice happens to be the two wavelengths we have already mentioned: 486nm in the blue, and 656nm in the red*. This dispersion is, in fact, the standard one for all glasses, and a simple 'V value' refers to this interval. If, now, we make up a pair of lenses whose focal lengths are inversely proportional to their V values, the combination will be achromatic for the two colours given above. The principle of the achromatic lens is as simple as that!

Let us take an example, and suppose that we want to calculate an achromatic combination using crown glass of N=1.52 and V=60.0, and flint of N=1.60 and V=36.0. The focal length of the finished lens is to be 150cm. We can find the focal lengths of the individual lenses from the formula

$$\frac{1}{F}=\frac{1}{f_1}+\frac{1}{f_2}$$

where F is the focal length of the combination and f_1 and f_2 are the focal lengths of the crown and flint elements (note that f_2 is in fact negative, since it is the focal length of a diverging lens). Knowing that $f_1/f_2=-36/60$, we can easily deduce that $f_1=60$cm and $f_2=-100$cm.

The radii of curvature of a lens of focal length F and refractive index N are given by

$$\frac{1}{F}=(N-1)\left(\frac{1}{r_1}-\frac{1}{r_2}\right)$$

*These apparently arbitrary values are chosen because they correspond to distinctive dark lines, known as the F and C lines respectively, that are observed in the spectrum of the sun.

where r_1 and r_2 are the radii of curvature of the first and second surfaces of the lens. Clearly, we can choose an infinite number of pairs of curves that will produce a given focal length. Let us take the very convenient case where the second surface of the crown lens has the same radius of curvature as the front of the flint, while the rear surface of the flint is flat (i.e. r_2 equals infinity). First of all, calculate r_1 for the flint. The item $1/r_2=0$, and so we have

$$\frac{1}{100} = 0.6 \times \frac{1}{r_1}$$

giving $1/r_1=-0.0167$, or $r_2=-60.0$cm. The negative sign means that the surface is concave to the incident light. We now also know r_2 for the crown, and so r_1 can be obtained from

$$\frac{1}{60} = 0.52\left(\frac{1}{r_1} + \frac{1}{60}\right)$$

$$\frac{1}{60} = 0.52\left(\frac{1}{r_1} + \frac{1}{60}\right)$$

which gives $r_1=65.1$cm. Numbering the surfaces through from 1 to 4, in the usual manner, we obtain the following prescription for our achromatic lens:

Crown: $N=1.52$, $V=60.0$
Flint: $N=1.60$, $V=36.0$
Focal length$=150$cm
$r_1=+65.1$cm
$r_2=-60.0$cm
$r_3=-60.0$cm
$r_4=$Infinity

Other aberrations

So, with the aid of elementary mathematics, it is possible to calculate the radii of curvature of an achromatic object glass given the glass types to be used. (The thickness of the lenses has little effect on the result; we would normally arrange for the thinnest part of each – the edge of the crown and the centre of the flint – to be about 1/20th of the diameter.) However, having achieved an achromatic pair, the designer would then want to check on the spherical aberration, adapting the curves if necessary, while he might also consider the elimination, or at least the moderation, of off-axis effects such as coma and astigmatism. This flexibility is another virtue of the object glass. A Newtonian mirror must be parabolic, and all its aberrations are circumscribed; but an object glass contains four separate curves. By manipulating these 'variables', and also choosing the appropriate glasses, it is possible to obtain objectives with different characteristics although of the same degree of achromatism. In particular, spherical aberration can be made so slight that relatively little figuring is required to make the objective perfect – many f/15 designs are, in any case, well within the Rayleigh limit for spherical aberration with simple spherical curves.

Types of object glass

Most amateur objectives are of the 'contact' form, with the two elements in physical contact (cemented) or separated by slivers of aluminium foil. Widely-separated object glasses are found mainly in large observatory instruments. Of the contact objectives, the most highly-corrected is the *Fraunhöfer* type (Fig. 30), called after its famous maker. Coma is completely absent, making it relatively insensitive to squaring-on adjustments, although astigmatism will still appear if the maladjustment is gross. Spherical aberration is also small, and negligible in normal apertures at f/15 – not that this counts

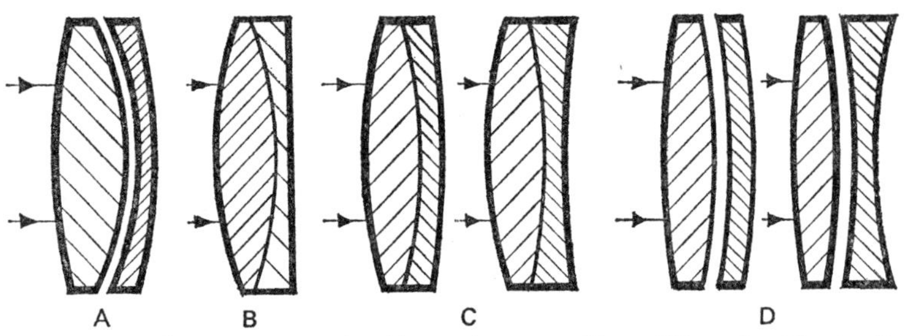

Fig. 30 Different types of object glass. A: Fraunhöfer. B: 'flat-back' contact type, recommended for amateur construction. C: Clairaut type, with second and third surfaces matching. D: Littrow type, with equi-convex crown lens. The general distribution of curvature within these families can vary a good deal

for much, when it can be removed so easily in any case. The Fraunhöfer type is particularly suitable for fast, wide-field instruments such as comet-seekers. To pay for these advantages, the lens is difficult to construct since all four surfaces have different radii of curvature. It is certainly not recommended for a first attempt.

The *Clairaut* type is very often used in binocular objectives. Here, the second and third surfaces match and the lenses can be cemented together provided they are not more than about 6cm in diameter – if larger, a hard-setting cement may twist them. The other simple type of objective is the *Littrow*, with an equiconvex crown component. Both these types suffer from coma, but visual fields at f/15 will not be affected.

Two 'flat-back' designs for the amateur

There is much to be said for manipulating the curves to give a plane surface on the rear of the flint, and several designs incorporating this feature have been published. A flat surface clearly requires less rough grinding than does a curve; and the flat tool, when polished, can be used as a test mirror in the way to be described. The two following designs, both by H. E. Dall, are very suitable for first attempts.

Design 1

Crown: $N=1.519$, $V=60.3$
Flint: $N=1.620$, $V=36.2$
$r_1=+0.421$
$r_2=-0.421$
$r_3=-0.421$
$r_4=\text{Infinity}$

Design 2

Crown: $N=1.524$, $V=59.0$
Flint: $N=1.620$, $V=36.2$
$r_1=+0.423$
$r_2=-0.385$
$r_3=-0.385$
$r_4=\text{Infinity}$

The radii of curvature are given in terms of the required focal length of the finished lens. Design 2 has the slight drawback over design 1 of requiring a separate glass tool for the first surface, instead of being able to grind both sides of the crown on the flint; but in the writer's opinion this disadvantage is less serious than it may appear, because it is not easy to maintain perfect grinding contact across two surfaces worked simultaneously on the same tool. Also, the crown glass used in design 2, which is known as *ophthalmic crown*, is very easily obtainable because it is manufactured in large amounts for making spectacle lenses; it also costs much less than special optical glass made in relatively small batches. It will, therefore, be assumed that design 2 is being attempted, even though design 1 has somewhat better spherical aberration. A 10cm f/15 object glass is envisaged.

The glass blanks

First of all, we need to know how thick the lens discs should be. We have

Crown: $r_1 = +63.45$cm, $r_2 = -57.75$cm
Flint: $r_3 = -57.75$cm, $r_4 =$ Infinity

The depths of curve (sagittae) come out at $r_1 = 2.2$mm, $r_2 = r_3 = 2.4$mm, and $r_4 = 0$. Hence, the optical thickness of the crown's centre is 4.6mm, and that of the flint's edge is 2.4mm. Adding the somewhat arbitrary 1/20th of the diameter, we obtain centre and edge thicknesses for the crown and flint lenses of 9.6mm and 7.4mm respectively. To allow for the considerable loss of general thickness involved in the roughing-out process, the blanks should be generously over these values: say 15mm and 12mm respectively. The supplier will advise the nearest convenient thicknesses.

The exact diameter of the lenses does not matter; 10.5cm will be ample to allow for the cell's retaining ring. However, the two components must be of the same size to within about 0.05mm, as otherwise they will not centre on each other sufficiently accurately when placed in the cell. Two circular glass tools of the same diameter, cut from 12.5mm plate, will also be needed.

Starting the grinding

Examine the two surfaces of the flint disc to see if there are any surface defects or *occlusions* caused by foreign matter sinking into the glass. These are a common feature of optical glass, and are another reason why a generous allowance should be made for thickness reduction during the grinding, since such defects will have to be ground out. Select the worst side for the concave face, as this will receive more grinding. Now, after bevelling all the edges, secure the crown disc to the stand and grind the flint face-down on it with grade 80 carborundum. Particular care must be taken to support the lower lens properly; if it rests on an uneven surface it can easily flex and acquire an irregular surface. The best course is to make up a plywood disc, with its surface roughly curved to accommodate the face of the disc being worked, and to secure the glass to it with a band of PVC tape. If a lining of thin foam plastic is placed between the lens and the backing, the disc will effectively 'float' free from strain.

Since a relatively steep curve has to be produced, side stroke is advisable. As the convex curve develops and the bevel narrows, renew it to avoid ugly chips down the side of the lens.

Making and using gauges

A very convenient way of checking the curvature of the lenses is to make a set of gauges. Gauges are impractical when mirrors of normal focal ratio are being made, but in the case of steep lens curves they produce more accurate results than splash tests. Take a length of wood and screw a glass-cutter with a sharp wheel to one end. Measure along from the wheel a distance equal to the radius of curvature required, drill a small hole, and tap a nail through. Hammer this nail into a board, so that the glass-cutter can be swung through a short arc. Now take a piece of window glass measuring about 11 x 5cm and secure it, by knocking tacks in around its edge, to the board under the cutter. Take *one* firm sweep from end to end of the glass. Place the glass, cut uppermost, on two small nuts, one at each end of the cut, and press down on the two long sides. If all is well, the glass will snap along the curved cut. This is much easier than it sounds, and curves of surprisingly short radius will snap cleanly if the wheel is new and the glass is not too thick. Grind the two arcs on each other with 320 carborundum until their faces are evenly grey, and bevel them until only a narrow strip of grey remains. These gauges are now accurately matched, and can be used for checking the work by seeing if the optical surface (more conveniently, the convex face of the pair) shows a light-tight fit. Make another pair of gauges for the first convex surface.

It is best to take the grinding of the second and third surfaces through to the 320 stage only. There is nothing to stop the work from being taken through to the finest emery, except that subsequent roughing operations on the first and fourth surfaces might result in damage from the heavy handling. At the conclusion of the 320 grade, as well as showing even surface quality all over, the curves should match their gauges perfectly. The dangers of poor contact and residual pits, and the advisability of regular reversing of the discs, have already been emphasised.

Removing 'wedge'

The first surface of the crown lens is now tackled, using its plate glass tool. It may be necessary to thin the lens down somewhat by extra grinding. Once the rough curve has been produced, checks must be started to examine the variation in edge thickness, which is generally known as 'wedge'. Wedge produces coma (gross wedge makes the lens act like a prism, and turns stars into short spectra!), and it must be ground out to the point where no difference of edge thickness greater than about 0.005mm can be detected. In commercial establishments producing small object glasses for binoculars and the like, lenses are de-wedged after polishing by centring them optically on a spindle and grinding any eccentricity off the edge, but this is wasteful of glass where larger objectives are concerned.

A simple jig (Plate 7) will tell all that needs to be told. A metal baseplate carries three steel balls marking out a circle somewhat smaller than the diameter of the lens component. Two vertical stops serve to locate the lens laterally, and between them a dial gauge, micrometer, or even a simple screwed rod with a steel ball at the end descends to make contact with the margin. Successive readings around the circumference indicate the variation in thickness. The writer uses a 2 BA rod, with a disc marked in divisions at the top, and this gives the required accuracy at a fraction of the cost of a purpose-built measuring instrument.

The way of reducing wedge is, of course, to bear extra heavily on the thick part of the lens. Such selective wear must be interspersed with periods of normal grinding to ensure that a revolutionary surface is maintained. If the remaining wedge is reduced to about 0.05mm on the 80 grade, the rest can be taken out at the 220 stage – the intermediate grade will then be able to take out any remaining unevenness from the 80 work. Checks must, of course, be maintained throughout the smoothing, to ensure that any fresh wedge, if introduced – which it should not be if the operator performs his revolutions correctly – is quickly eliminated. Some makers, rather then spend time in this admittedly tedious work, resign themselves to having imperfect elements and rotate them against each other during testing to find the best mutual orientation, when the wedges in the elements more or less cancel each other out! The chances of true cancellation in this way are remote, and all that is usually obtained is a minimum-error position. Such sloppy work is invariably characterised by pencil marks on the edges of the two lenses to indicate the orientation required. A good lens, made to the tolerance advised, is utterly insensitive to mutual rotation of the components.

Bringing and holding the curves to truth, and eliminating

edge thickness errors at the same time, is not an easy task, but it is one that must be done well if the finished objective is to be successful. If the curves wander perceptibly from their gauges, the focal length, spherical aberration, and chromatic correction will all be changed to some extent. The accuracy of the first and fourth surfaces is more critical than that of the second and third, which tend to compensate each other. For example, if the radius of curvature of the first surface is too short, then the focal length of the crown element will be too short, making the flint lens effectively too weak – hence the lens will appear under-corrected for chromatic aberration. If the intermediate radii are too short, the focal lengths of both the crown and flint elements will be reduced in approximate proportion, and the colour correction will remain about the same.

The first curve is taken right through to the finest emery, and the lens is put aside. The other plate glass tool is now taken up, and the flat side of the flint lens is tackled with carborundum and wedge-tester.

Checking the flat surface

It is possible to bring the flat surface within the limits of accuracy required by the design by checking it against the light with a straight-edge. This is, after all, how we compare the other surfaces with their gauges. But it is easy to obtain a more quantitative result by making a simple *spherometer,* and this accessory will be found useful for checking other pieces of glass to see if they are approximately flat. The underside of a metal disc carries three steel balls that will comfortably stand on the lens or tool. Through the centre of the disc passes a gauge – a screwed rod, as described above, will serve – and this gauge is set to contact on the lens and the reading noted. It is then set to contact on the tool. If lens and tool are in contact all over, as they should be, the difference between the readings will represent *twice* the sagitta of each surface; if the readings are the same, then both must be flat. This is a most sensitive test, since the errors are effectively doubled. If the spherometer can read to a difference of 0.005mm, which is possible with a 2 BA rod, the maximum possible departure of each surface from true flatness can be only 0.0025mm or 5 wavelengths of yellow light. Such flatness is quite accurate enough for the purpose.

It is possible that carborundum no coarser than 120 or even 220 will smooth out the initial irregularities on the flint's surface. Before deciding which grade to use, however, check the wedge and decide, on the basis of the experience gained on the crown lens, which abrasive will do the quickest job. Remember also that flint glass grinds away much more quickly than does crown. In fact, flint grinds decidedly coarsely – it is very brittle, and fractures more readily – and it is advisable, in the interests of quick polishing, to use the finest possible abrasive in the final smoothing. A 125 finish will take a considerable time to polish. Object glasses generally have a reputation for being somewhat slow to polish compared with mirrors, the reason being that their small size makes it difficult for the operator to grind heavily and break down the abrasive into the fine particles that give such a smooth surface at the end of a wet. The answer is to use very fine abrasives to begin with. A few final wets of 95 or, preferably, 50 aloxite will make a very significant difference to the polishing time, particularly as far as the flint component is concerned.

Polishing the lenses

It will probably be best to polish the flint component first, since the polisher for the concave side must be formed on the second surface of the crown lens. Great care must be taken in cutting the channels to avoid scratching the fine-ground surface with the blade. The flat polisher can be formed on the back of the concave plate glass tool, and this can also be used to give the flat tool itself a fair polish, since it will be needed for testing. Both sides of the crown lens can be polished using the concave tool as a former for the pitch; it will not matter that its curve does not exactly match the radius of the second surface, because the pitch can easily be pressed into shape. The technique of polishing does not differ in any way from that already described for a mirror.

Testing the object glass

The classic way of testing an object glass is by the *auto-collimation* method shown in Fig. 31. The lens is set up

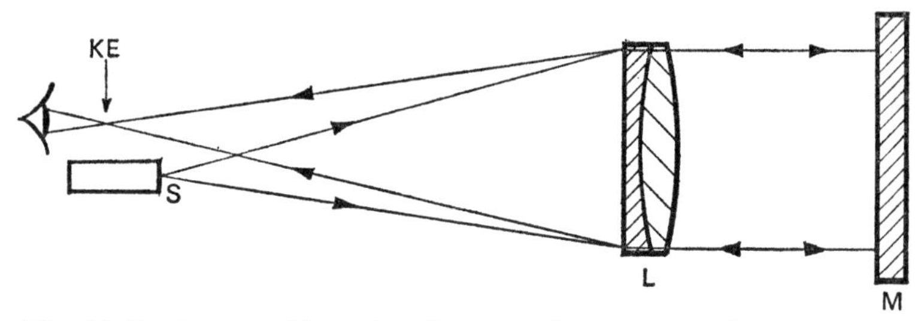

Fig. 31 Testing an object glass by autocollimation. L: object glass. M: plane mirror. S: light source. KE: knife-edge

facing a plane mirror, and an artificial star and knife-edge are situated at its focus. If the lens is perfect, it produces parallel light from its front surface which is reflected off the mirror and back through the lens again, forming a true image of the artificial star and showing even darkening when the knife-edge is moved across. Imperfections of figure have a doubled effect since the beam passes through the lens twice, and zones and residual aberration too small to be visible when the lens is used in the telescope can be detected with certainty. This super-accurate test, coupled with the relative insensitivity of refracting surfaces to errors of figure, greatly simplify the optician's task.

It is often thought – and this may be a reason for the wide prejudice against amateur object-glass making – that the test flat mirror used in the autocollimation test must be of super-fine quality. This is not so. Certainly it should be free from major zones and asphericity, but these will not occur if the correct polishing technique is adopted. In any case, faults can be detected during the testing itself, and either treated or allowed for. Overall flatness, which is much harder to attain without special tests, matters little. For example, if the test mirror is 5 wavelengths of light convex – an amount just detectable with the simple spherometer already described – all that happens is that the object glass, instead of being tested at the infinity focus, is effectively being tested on an object about 500 metres away. This corresponds to 'infinity' for all practical purposes. In fact the only time when flat mirrors really do have to be accurately plane, rather than spheres of very long radius (such as the 500 metres in the above example), is when they are used at appreciable angles to the direction of the light striking them – as in a telescope diagonal – when sphericity of the order of a wavelength can cause the reflected beam to be appreciably astigmatic.

The check for mirror zones is straightforward. If a zone appears when the test is set up, move the mirror a little way laterally so that an arc of its margin shows in the lens aperture. It will now be easy to see to which circumference – that of the lens, or that of the mirror – the zone is concentric. This procedure is, in any case, desirable to check whether the margin of the flat mirror is turned off. As a general rule, when using test mirrors or collimators of the same aperture as the work being checked, the two optical apertures should be left slightly offset to prevent confusion over marginal residual errors of figure. Vertical offsetting is best, since the sweep of the knife-edge shadow will still traverse almost a full diameter of the object glass.

The test mirror need not be coated. It need not even be perfectly polished; a 30-minute spell will shine up enough of the surface to give ample illumination with a bright pinhole. What is important, however, is for the optical axis of the object glass to be accurately in the normal of the test mirror. If it is tilted, the rays will be passing through the object glass at an angle, and the result will be that we are observing the off-axis performance of the lens, which almost certainly means a comatic image. The easiest way of achieving correct orientation is to lay the test mirror face-up on the floor or a horizontal surface, and place the object glass face down on top of it, testing vertically. If this is found to be inconvenient, a simple holder can be made up to hold the lens square-on to the mirror for horizontal testing.

Although final aspherising can be carried out on any (or all) of the four surfaces, it is more convenient to work on one of the external surfaces, since the two lenses can be held permanently together with a length of waterproof tape around their edges. Three small pieces of cooking foil are set at 120° intervals around the margin to hold the surfaces slightly apart. The fourth surface is the best one to work during the final figuring, because the convergence of the light in the object glass means that the *effective* rear aperture of the lens is less than that of the front, and any turned edge imparted during the work may lie outside the working area. Turn-off at the extreme margin will probably be masked off by the retaining rim of the cell.

Knife-edge testing

A lens with well-polished surface will show a regular, slightly undulating figure. The first thing to remember is that the figure of an object glass appears *reversed* with respect to that of a mirror. The easiest way of accustoming oneself to the transformation is to imagine that the knife edge is travelling in the opposite direction, so that a hill looks like a hole, etc. The reason for the change is that, in the case of a mirror, a depressed region has a shorter focal length because it converges the light more strongly; a depression on a lens has a diverging effect, and hence a longer focal length. If the object-glass designer has produced, within the limits of the glass available, a truly corrected system, the general cut-off will be flat but with a slight 'ripple'. This ripple represents the higher-order spherical aberration that cannot be removed by using simple spherical surfaces, even when they are distributed over the four faces of an object glass. An eyepiece test of such a lens will show approximately equal expanded discs containing anomalous light and dark zones. Alternatively, there may be slight general under- or over-correction caused by some asphericity imparted to one or more of the surfaces during polishing. If the aberration is no more than a few millimetres, it can be figured out using a small soft polisher, while very slight zonal irregularities are best worked on with the thumb or the heel of the hand.

The colour correction

This will also be well displayed by the knife-edge test. As the knife-edge passes across near the focus, the object glass is seen to be divided into two halves of different colour, showing the tints for which it has extreme foci. Remembering that we have brought the red and blue together, with the yellow focusing somewhat shorter, we see the red and blue extremity of our correction – a Victoria plum colour – contrasted with the yellow-green of the minimum focus. Makers commonly call this effect 'plum and apple', and a very welcome sight it is too, after days or weeks of patient grinding and polishing in the hope that the curves are right! Of course, there are different degrees of achromatism, and the tints to be seen, or expected, depend upon the colours selected for coincidence and minimum focus by the designer. Some makers have used designs with the minimum focus closer to 'pure' yellow, say 565nm, rather than the 555nm commonly used, while many modern designs have pushed the minimum focus well into the yellow-green. It is mostly a matter of individual taste. For photographic use, however, there is an important difference, for photographic emulsions are normally mainly sensitive to light at the blue end of the spectrum. Photographic objectives therefore bring the minimum focus near 490nm, with yellow and blue-violet light combined just behind; the red is left well back, as is the violet in a visual combination. Testing a photographic objective would therefore show bright green instead of plum colour, and blue instead of yellow-green.

It is worth noting that if we combine three elements instead of two, coincidence of focus for three colours is possible, and the degree of secondary spectrum is reduced to a low level (of the order of 1/10th that of a doublet). Alternatively, a triplet objective can be used to produce coincidence of visual and photographic foci, earning the name *photovisual.* In practice, because of their extreme sensitivity to misalignment in the cell, the use of triplets is confined mostly to relatively small lenses of short focal ratio, where the outstanding purple halo becomes obtrusive.

Identifying faulty surfaces

Should the assembled lens show a severely zoned or aspheric appearance, one or more of the surfaces has been worked badly. It is far better to re-polish the offending side than to try to remove large quantities of glass arbitrarily from the rear of the flint by local polishing because, since a lens is only a quarter as sensitive to errors of figure as a reflecting surface, that much more work must be done to remove faults! It is also a more workmanlike way of going about the job. This is where the maker will find himself in some difficulty, because only one surface, the concave side of the flint, can be tested directly with the Foucault apparatus to see its figure. At any rate, this is the first surface to examine. If it is spherical, or nearly so, it is possible to eliminate the second and third surfaces by setting the objective up on test with the foil spacers removed and a saturated solution of sugar between the elements. The refractive index of this solution is about that of crown glass, so we have eliminated the short but sharp jump that the light suffers on passing through the narrow air gap (refractive index 1.0) between the two elements. Instead of a difference of about 0.5, the difference is now only that between the refractive indices of the two glasses used, or about 0.1. This means that any zones on the second surface will be almost eliminated, and the test should show an improved appearance.

An objective with an equiconvex crown lens can have the component reversed and the first surface also annulled by the same technique. Otherwise, there is no easy way of deciding which of the outer surfaces is at fault, and the only solution is to give one of them a 30-minute spell on a well-pressed polisher to see if any improvement is noticed.

Interference testing

A professional optician would have no difficulty in identifying the faulty surface. If two closely-matching surfaces are placed together and viewed in the light of one wavelength (known as *monochromatic* light), or at least of a very limited spectral range, dark and bright bands are seen across the surfaces. These *interference fringes* represent contour lines; if they are straight, the surfaces have the same radius of curvature, and if they appear as rings, one surface is either convex or concave with respect to the other. The distance between each dark or bright line corresponds to a distance between the surfaces of half the wavelength of the light source being used. By placing the plane side of the flint lens on a master flat, any asphericity would show up as irregular waves in the fringes. Similarly, placing the second and third surfaces on each other would betray departures from sphericity, and offsetting the lenses would indicate to which surface the errors belonged. Eliminating these three surfaces would establish the offender. Small sodium lamps, which form a convenient monochromatic source, are not expensive considering their usefulness, and the serious amateur optician would do well to acquire one, particularly if he intends making object glasses or flat mirrors.

Distortion

If the objective shows a persistent comatic or astigmatic appearance on test, no matter how it is adjusted relative to the flat mirror, it means one of several things. The two components could be at an angle to each other, or one or both could have significant wedge – both faults will give a comatic image. Astigmatism suggests that one of the lenses was deformed during polishing and has a 'bent' surface. This can very easily happen to a face-up lens during polishing, if it is not supported on a resilient backing as advocated above. To identify the faulty component, rotate one lens with respect to the other, and observe how the axis of the coma or astigmatism rotates. Prevention of either of these disasters is infinitely better than the cure!

The cell

An object glass should really be mounted in a metal cell turned in a lathe to accept the discs with just enough space to give a slight rattle when shaken from side to side. A lip left against the front surface serves to locate the objective, which is held in by a ring, once again with a certain amount of play (Fig. 32). Anyone without turning facilities can produce a cell that is equally effective, though less elegant, by cutting a hole in a disc of thick plywood and retaining the taped-up objective therein by securing two thin plywood discs, cut out to the required aperture, on either side of the holder. The cell, whatever its nature, must be mounted at the end of

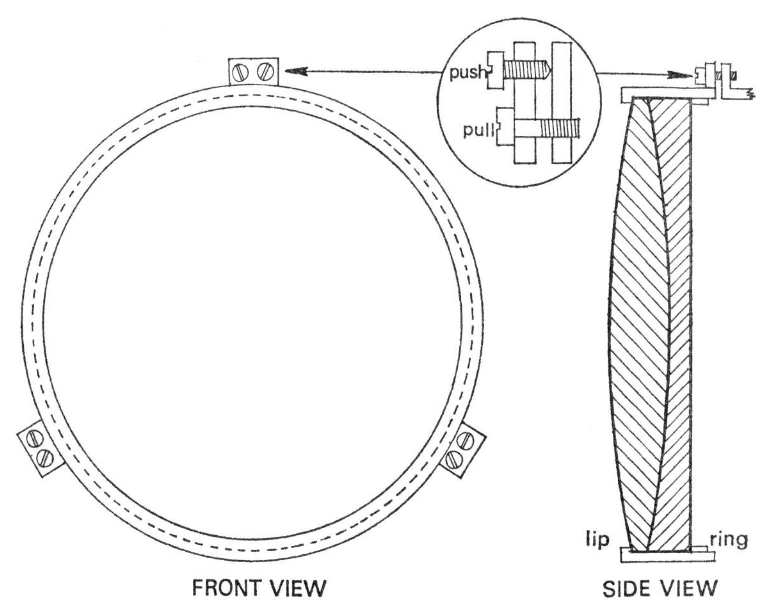

Fig. 32 A simple cell for an object glass. The action of the adjusting screws linking the cell to the tube is shown

the telescope tube with push-pull screws to allow precise squaring-on.

Eliminating the air space

Although it is common to assemble Clairaut-type objectives with foil spacers, it is far better to eliminate the optical gap between the lenses altogether. As we have seen, hard cement of the type used in binocular objectives is liable to twist larger lenses, or to break up due to the different coefficients of expansion of crown and flint glass. An air space, however, results in the loss of some 8 per cent of light through the two air-glass reflections; to this must be added the possible reduced contrast caused by some of the scattered light being re-focused near the image. The effect can be significantly reduced by blooming, but this is relatively expensive. A better alternative is to place a few drops of microscope immersion oil, having a refractive index intermediate between that of the two elements, on the centre of the concave surface and to lower the crown on top. Press out the excess oil, wipe the edges as clean as they will go, and seal them with a tight ring of polythene cut from a small bag. The oil will never dry out, and the lens will act optically as if it were made from one piece of glass. Figuring should be carried out with the objective oiled because of the ameliorating effect on zones, as mentioned above.

Aligning the object glass

The squaring-on of an objective is not a difficult matter, and if the cell is properly made the adjustments should be permanent. An easy way of obtaining approximate alignment is to cover the objective and to view the reflection of a torch bulb held at the open end of the drawtube, with the eye close behind and as near as possible to the axis of the tube. An uncemented objective will show four images of the bulb, while an oiled one will show only two, from the first and fourth surfaces. If the objective is truly adjusted, these images will be superimposed on each other when both lamp and eye are central. Once this position is achieved, observation of a star in the centre of the field will indicate, by the orientation of its comatic flare, the adjusting screw to be turned. Once obvious coma, if any, is eliminated, slight residual misalignment is revealed by non-uniformity in the diffraction rings of the slightly-expanded discs.

The star test

An objective figured to give a clean cut-off in the autocollimation test will give breathtaking star images. Here we find, if the seeing conditions are reasonable, the true 'disc and rings' of the textbooks, the fruits of a highly accurate test, lack of central obstruction due to a diagonal and its supports, and a long focal ratio. The purple halo of secondary spectrum will be barely detectable except on the planets Venus and Jupiter, and the colours in the expanded discs will appear far less obtrusive than might have been expected from the autocollimation appearance. At the best focus, a white star will appear truly white, or with the faintest trace of yellow. A reddish halo, which expands as the eyepiece is moved inside the focus, indicates under-correction; a well-corrected object glass will show just a trace of red in the margin of the intra-focal disc, and perhaps a small reddish point at the centre of the extra-focal disc, marking the focal point of the red rays as somewhat beyond that of the yellow. In the objective described above, the difference between the red-blue and yellow foci should be about 0.6mm. The margin of the extra-focal image will have a yellow-green fringe.

The tube and mounting

Refractor tubes used always to be made of brass, but the high price of the metal today makes it something of a luxury, although it can always be argued that a fine lens deserves to have a few extra pounds spent on it. PVC is popular, but is relatively heavy, is rarely straight, and is flexible. Seamed or seamless galvanised or light alloy tubing is probably the best compromise. A couple of 'stops' should be fitted at intervals down the tube to block out any reflections off the inside of the tube from bright objects near the telescope's line of sight; in effect, they mark the diameters of two points along a truncated cone formed by the objective at one end and the field stop of a low-power eyepiece at the other. The sizes of the holes can be calculated from the formula for the minor axis of a Newtonian diagonal given on page 128.

Some principles concerning the mounting of refracting telescopes were outlined in Chapter Two. Because of the length of the tube, and the desirability of having a portable unit, the writer would advocate repeating the two-point tube support design. A fork and trunnions can replace the simple block and hinge at the top of the tripod, and telescopic or sliding tubes will give more rigid support than cord. These are matters that the amateur who has already made a simple refractor and a Newtonian telescope will readily solve for himself.

CHAPTER TEN

Larger Reflectors, and Advanced Telescopes

Although it has been assumed, in the previous chapters on mirror-making, that a 15cm mirror is being attempted, the choice of aperture for the beginner's first reflecting telescope is a matter for his own judgment. Some writers have asserted that, since a 20cm telescope will collect almost twice as much light as a 15cm model without being twice as difficult to make, it therefore represents a sounder proposition. The validity of this argument depends on the meaning of the world 'difficult'. Many people will find a 15cm mirror quite difficult enough. It does not necessarily follow from this that they would have found a 20cm impossible, but it undoubtedly represents a longer job, and, since few people would deny that a large mirror, especially if its focal ratio is relatively short, is a harder proposition, the chances of achieving first-class images must be smaller. Since a 15cm telescope will give vastly better views than the binoculars or cheap refractor that the beginner may initially possess, he is unlikely, in any case, to appreciate fully the benefit of the extra aperture over which he will have had to toil.

Making a larger mirror

The enthusiast, having made a small mirror and finding optical work to his liking, will undoubtedly soon be thinking in terms of another project. It is to be hoped that he will follow the advice of the previous chapter, and make an object glass; but subsequently he will certainly be envisaging a much larger aperture, and this must mean a reflecting telescope. A 25cm or 30cm mirror might then be undertaken.

The aspect to be considered most critically is focal ratio, for this affects, more than mere aperture, the difficulty of figuring a mirror. Figuring resolves itself as the problem of rendering a spherical mirror aspheric, and the formula already given, $\delta S = D/50 \times \left(\frac{10}{f/}\right)^3$ wavelengths of yellow light, tells us that the depth of glass to be removed from parabolic mirrors of various sizes is proportional to the aperture but inversely proportional to the *cube* of the focal ratio. In other words, there is twice as much figuring needed on an f/8 mirror as on an f/10; a table representing the depths of glass to be removed at different sizes and focal ratios follows.

THE AMOUNT OF GLASS TO BE REMOVED IN PARABOLISING A MIRROR

	Focal ratio						
Aperture (cm)	12	10	8	6	5	4	3
10	0.12	0.20	0.39	0.92	1.6	3.1	7.4
12	0.14	0.24	0.47	1.1	1.9	3.7	8.9
15	0.17	0.30	0.60	1.4	2.4	4.7	11
20	0.23	0.40	0.78	1.8	3.2	6.2	15
25	0.29	0.50	0.98	2.3	4.0	7.8	19
30	0.35	0.60	1.2	2.8	4.8	9.4	22
40	0.46	0.80	1.6	3.7	6.4	12	30
50	0.58	1.0	2.0	4.6	8.0	16	37

These amounts represent the approximate difference in central depth, in terms of wavelengths of yellow light, between spherical and parabolic mirrors of the size stated.

The extra *volume* of glass to be removed from large as opposed to small mirrors of the same focal ratio does not count for much, at least in amateur sizes. Large mirrors tend to aspherise at a reasonable speed because of the longer and heavier strokes that can be applied during small-polisher aspherising. What is difficult is not so much to remove glass in large amounts as to take it off adjacent zones in significantly different quantities, which has to be done when steep paraboloids are attempted. To sum up, it may well be found as difficult to figure a 15cm f/6 mirror as a 30cm f/8; therefore, the enthusiast would be well advised to follow his 'long' 15cm with a 'short' one of perhaps f/5. This will also provide him with a useful, highly-portable and powerful telescope for taking on trips and holidays. The amateur who can make such a mirror giving fine star images can consider himself out of the apprentice stage, and larger apertures can be tackled with confidence.

The extra work involved in grinding and smoothing large mirrors is of less account than might be thought. Certainly, hogging out a 30cm disc is toilsome, but large discs smooth at the same rate as small ones. In the final stage of smoothing, their extra weight breaks down the emery very quickly, and an extra-fine surface may be achieved that will polish *more quickly* than a smaller mirror might.

Tests for aspheric surfaces

A critical factor in aspherising larger or steeper optical surfaces is the sensitivity of the test available, and it must be admitted that the simple Foucault test is not well suited to the production of a surface departing by several wavelengths of light from a sphere. In such cases the knife-edge movement is large and the mirror's surface shows a highly-contrasted set of shadows in which significant zonal errors can lurk unseen unless sought by an experienced eye. Fast paraboloids usually appear pleasingly smooth and regular under the knife-edge, and the unsuspecting worker may congratulate himself upon his excellent technique until he 'knifes' his mirror on a star or in an optical null test and discovers all sorts of ugly hills and hollows that lay submerged in his 'doughnut'. Moreover, it is no longer possible to interpolate across three zones only, as is the case with an f/10 or f/8 mirror. To establish the overall shape, perhaps five or seven zones must be checked, with the possibility of slight errors of shadow-judgment allowing significant zonal errors to pass. The writer is convinced that the poor shape of many amateur mirrors is due less to lack of skill in treating errors than to the difficulty of interpreting what shape the surface really has. An outline of more precise mirror-testing methods is given in the hope that it will enable ambitious amateurs to give themselves a proper chance of achieving a fine surface.

Testing by collimated light

The supreme test of a mirror should be the star-test; for reasons already outlined, this is largely impractical. Simulated starlight – parallel light – can, however, be produced in the optical workshop by using a 'collimator', a mirror or lens of known excellence which has an illuminated pinhole placed at its focus. The system then emits parallel light, which is brought to a focus again by the mirror or lens under test (Fig. 33). For full illumination, the collimator objective should

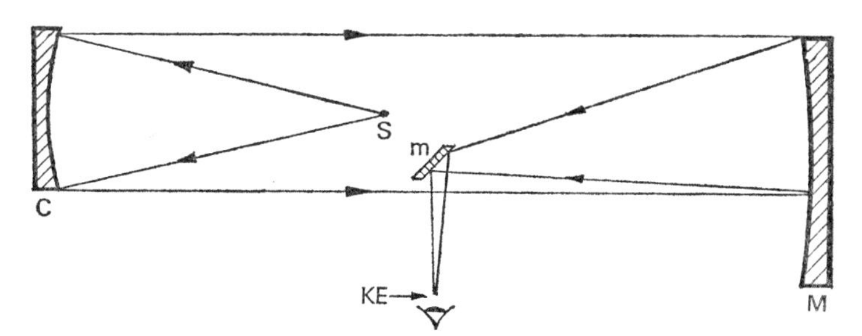

Fig. 33 Testing a mirror by collimated light. C: collimator mirror. M: mirror being tested. m: flat mirror. S: light source. KE: knife-edge. The drawing shows a collimator being used to test a mirror of considerably more than its own aperture; note the slight marginal offsetting

be as large as the work being checked, but it is perfectly possible to use an offset collimator of half the aperture, set up so that the condition across a radius of the work can be checked. Bearing in mind the importance of the marginal regions, an overlap should be arranged so as to show a significant arc of the work's edge; the very centre is unimportant, since it is covered by the diagonal. Ideally, however, the collimator should be larger than the work, for any residual aberration in the collimator objective is reduced in importance as the squares of the two apertures.

The focal length of the collimator should be as long as possible, and in any case should be not less than that of the work, since the accuracy is reduced in proportion to the relative focal lengths. A 15cm f/10 mirror which has been proved on the stars could therefore be used to make a shorter-focus mirror of the same aperture, or up to a 30cm f/5 paraboloid, with results almost certainly better than those obtainable with the Foucault test.

Autocollimation test

So called because the test mirror is used to produce its own parallel light, which is then reflected back off a large plane mirror and re-focused near the pinhole (Fig. 34). This is essentially the same test as that used for an object glass, and it is the most accurate visual test for a parabolic mirror since

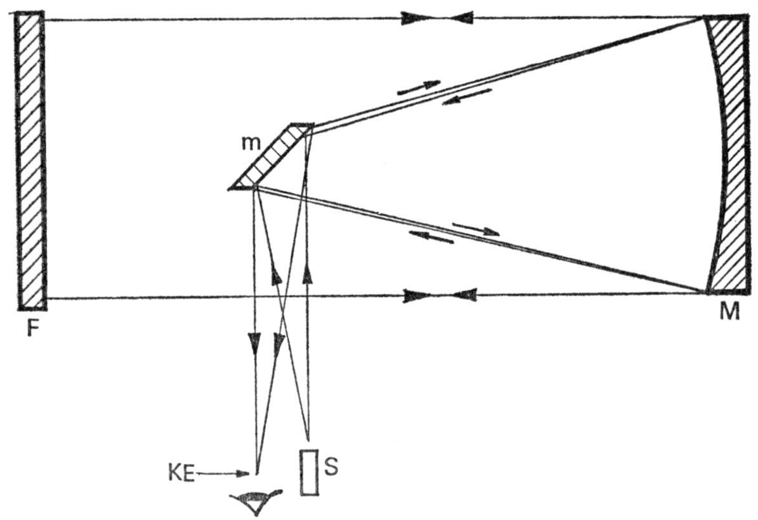

Fig. 34 Testing a mirror by autocollimation. F: plane mirror. M: mirror being tested. m: flat mirror. S: light source. KE: knife-edge. As with ordinary collimation, an undersize flat mirror can be used

the surface reflects the light twice, making defects stand out particularly well. However, compared with the use of the test in object-glass manufacture, the autocollimation method is an inconvenient one for paraboloids. The test mirror must be aluminised, and even then we have two reflections off the uncoated mirror, giving a relatively faint image. It is also a difficult test to set up because the alignment is so critical. In practice, it is doubtful if the method is more accurate than the null test described below.

Dall null test

This convenient test for a parabolic mirror, which can be arranged to suit prolate ellipsoids and hyperboloids as well, was first described by H. E. Dall in 1947. The general set-up is similar to the Foucault test except that the light from the pinhole passes through a plano-convex lens which imparts sufficient spherical aberration to cancel out that of the paraboloid. Hence a spherical mirror appears under-corrected (oblate) with an aberration equal to that of the required paraboloid. The optician then aspherises the surface until the central hill disappears and the mirror appears flat, at which point it must be a true paraboloid. This test is therefore as effective as having a collimator, and is capable of parabolising mirrors down to about f/4, in apertures of up to about 50cm, if a single lens is used.

It is curious that this null test has received so little attention; amateurs toil away reading deep Foucault shadows, with all the uncertainty that this process involves, when a couple of evenings' work could provide them with the means of producing far superior surfaces. A full account has been given by Dall (*Amateur Telescope Making*, Vol. 3, page 149), but sufficient information is given here in the hope that readers attempting more advanced mirrors will avail themselves of this test.

A lens, like a mirror, possesses spherical aberration, the amount depending on its distance from the object and the shape of the lens. The Dall null test is designed to be used with a plano-convex lens of normal crown glass, the focal length of which should be between 1/5th and 1/20th of the focal length of the mirror being tested. A lens of about 15cm focal length and 3cm diameter will test most mirrors the amateur is likely to make. Normal commercial quality will be sufficiently good for the purpose. The focal length (f) is found as accurately as possible by catching the image of the sun, or of a very distant street lamp, on a screen, and measuring the distance between the centre of the convex face of the lens and the screen. The flat side of the lens should face the object. There will be considerable blurring of the image due to spherical aberration, and the *maximum* distance at which a sharp image is obtained indicates the focal length required. The distance from the pinhole to the lens, in order for the test to match any paraboloid, is obtained from the graph in Fig. 35.

To make the tester, mount the lens, flat side outwards, at one end of a tube of suitable length, and have the illuminated pinhole unit in a short tube sliding into the other end. This enables the lens-pinhole distance B to be varied within the desired limits. A table can be prepared in advance, showing the position of the pinhole unit for different values of the mirror's focal length F. The only other feature required is a red gelatine filter between the lamp and pinhole to overcome the lack of achromatism of the single lens. It is quite easy to set B to within an accuracy of 1 per cent, and this is sufficient.

When setting up the test, the tube will be found to be very sensitive to alignment with the centre of the mirror, and any slight inclination will give a comatic image. Squaring-on is best done by using an eyepiece to observe the expanded image of the pinhole, and adjusting the tester until the discs are circular.

If, for any reason, it is desired to make a mirror that is under- or over-corrected, the setting of B obtained from the graph is multiplied by the appropriate value of *e*.

Let us now consider the other reflecting and mixed optical systems that the amateur can tackle.

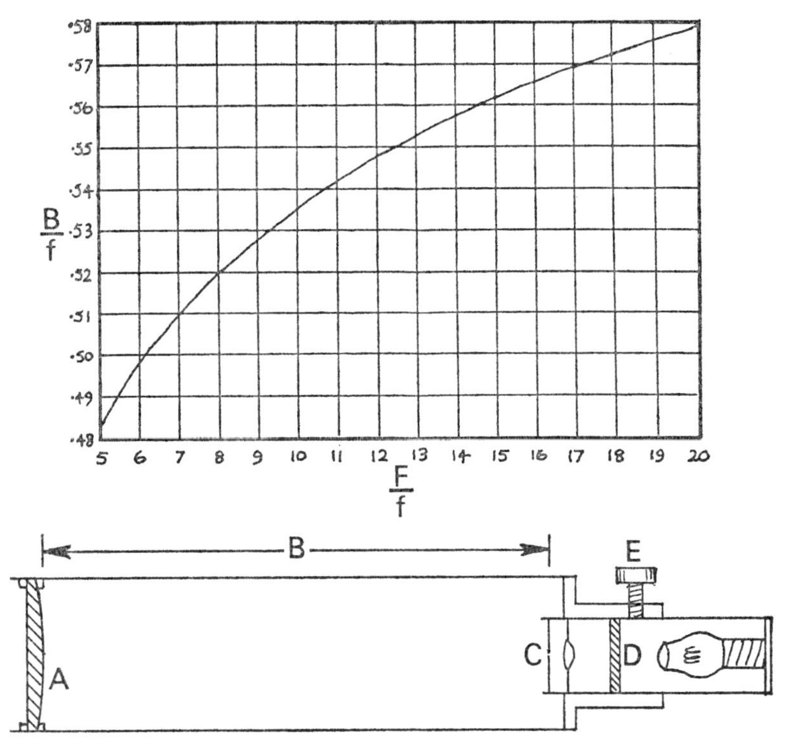

Fig. 35 The Dall null test. The upper diagram allows the lens-pinhole setting B to be determined from the focal lengths of the mirror being tested (F) and the tester's lens (f). In the lower diagram, A is the plano-convex lens, C is the illuminated pinhole unit shown in Fig. 15, with the addition of a red filter D, and E is a locking screw for use when the distance B has been set

The Cassegrain telescope

This system was mentioned in Chapter One (Fig. 8). It is the most compact kind of telescope known, and exists in many forms. Few amateurs have attempted pure Cassegrains because of the difficulty of figuring the hyperbolic secondary mirror, whose eccentricity is equal to $(A+1)/(A-1)$, A being the amplification afforded by the secondary. Thus, in an f/4-f/20 Cassegrain telescope (meaning an f/4 primary given an f/20 effective focal ratio by the secondary), the amplification is 5 and the eccentricity of the secondary would be 1.67. Achieving this relatively steep asphericity on a small mirror is not easy, but more serious is the lack of a suitable test, since a convex mirror cannot produce a real image of a point source. A number of ways of testing a prolate convex surface have been proposed, none of them entirely satisfactory. A far better solution is to leave the secondary spherical, and to use it in this form.

If this is done, we find that the curve on the primary mirror needs to be left somewhat under-corrected to give a spherically corrected overall system, the degree of eccentricity of the prolate ellipsoid being about 0.9, depending on the design. This means that less figuring is required in this so-called *Dall-Kirkham* system. With a null tester, the eccentricity can be set in advance, but another way of figuring a prolate ellipsoid is by the method of 'conjugates'. Any lens or mirror will give zero or minimum spherical aberration only when the object and image distances correspond to certain values. For example, a parabolic mirror gives an aberration-free image only when the object is at infinity and the image is at a distance equal to the focal length. Light being reversible, we could also speak of the object being at the focus, and the image being formed at infinity. These are the two conjugates of a parabolic mirror. The conjugates of a prolate ellipse can easily be calculated; if R is the mean radius of curvature, and e is the eccentricity, the positions of zero spherical aberration for object and image are obtained from $A=R/(1+e)$ and $B=R/(1-e)$. So, if we wish to test a 20cm f/4 ellipsoid of 0.9 eccentricity, the conjugate points A and B will be at distances of 84.2ccm and 1600cm from the mirror. An artificial star set up at the long conjugate can therefore be imaged, via a diagonal mirror, at the short conjugate, and the mirror figured until a flat cut-off is obtained.

If a source of monochromatic light is available, the tool on which the convex secondary was smoothed can be partly polished, figured spherical by the Foucault test, and used to check the shape of the secondary itself by observing the interference fringes. Alternatively, the optical components can be set up in their tube or a temporary rig, with the primary coated, and final figuring done on the secondary either in a collimator or using a star. All necessary formulae for calculating Cassegrain-type telescopes are given in Appendix B. In general the f/4-f/20 type would be recommended, with a secondary mirror about 2/9 of the diameter of the primary. Low-amplification systems result in over-large secondaries (for best results, the central obstruction in any telescope should not exceed a quarter of the aperture, unless there are overwhelming reasons for it to do so), and very high-amplification designs require a very short-focus primary if the final focal ratio is not to be excessively high.

Altering the eccentricities of the primary and secondary mirrors changes the off-axis performance of these telescopes. The pure Cassegrain has the same coma and astigmatism as would a parabolic mirror of the same effective focal length. The Dall-Kirkham design has inferior off-axis performance, but this does not matter with visual fields, particularly since the Cassegrain type is essentially a medium- to high-power telescope. It is even possible to have a spherical primary with a severely oblate secondary, but the field of good definition is so restricted that alignment is a major problem. By making the primary hyperbolic and the secondary still more so, the *Ritchey-Chrétien* configuration is achieved, with negligible coma over a wide field. This system is commonly used in large professional photographic instruments.

The Gregorian telescope

This system has a small concave secondary placed beyond the focus, instead of a convex one inside the focus. Since it is considerably longer than the equivalent Cassegrain, the major point to recommend it is that the secondary can be figured to a prolate shape by conventional methods. It also gives an erect image, which is useful for terrestrial viewing, but can be disconcerting to habitual users of ordinary astronomical telescopes!

Perforating a mirror

Another potential deterrent to the construction of a Cassegrain telescope is the need to perforate the primary mirror. It might be mentioned, in passing, that some Cassegrains have a small diagonal near the mirror to reflect the light from the secondary through a hole cut in the tube – the *Nasmyth* arrangement – but this involves an undesirable extra reflection. Drilling a hole through a glass disc is not as difficult as it sounds. All that is needed is a 'trepan' made from a piece of thin-walled metal tube of the appropriate diameter, with a shaft at one end to fit into a drill chuck. A bench drill set to a slow (300–400 rpm) speed is ideal, but the writer has drilled discs up to 25mm thick with an ordinary hand drill set in a crude jig with a slow feed facility. A vertical set-up is preferable, since the abrasive stays around the hole, where it is wanted. If the end of the trepan is fed with 120 carborundum and water or turpentine (which is less volatile, and gives a better cutting action), it will steadily grind its way through the glass, taking perhaps 1–1½ hours to do so if a 15cm disc is being tackled. Filing notches in the end of the trepan seems to aid the process. (See Plate 8.)

Glass is always in a state of strain, and it is no use trepanning a figured mirror. Even if the glass is perfectly annealed – as plate glass is – the surface is in tension, and it will be found that the region around a freshly-cut hole has risen up, causing a ridge of the order of a wavelength high. Poorly-annealed glass is strained throughout its depth, and trepanning causes a shock that can twist the whole disc optically and utterly ruin any figure that was on it; in discs larger than 20cm there is a very real risk of the glass breaking, and the disc must be fine-annealed beforehand. Some people drill the hole out right at the beginning, either working the mirror with the hole open or else cementing the core back in place with plaster of paris, so as to obtain a continuous optical surface throughout the optical work. This is all right, as long as no foreign matter is trapped into the gap between the mirror and core and released during the fine smoothing or polishing; sealing the join with pitch is a good preventative. Another way is to drill perhaps three-quarters of the way through the disc from the front, fill the groove with pitch, and finish the trepanning after figuring has been concluded.

The shade tube

Since the eyepiece in a Cassegrain-type telescope is facing the object, light from the sky around the secondary can fall on the field lens and flood the view, or at least severely lower the contrast, if the telescope is being used in twilight, while the instrument is useless in the daytime unless equipped with a *shade tube.* Much has been made of the difficulty of eliminating sky-flooding, and many people must have been deterred from making a Cassegrain because of the apparent magnitude of the problem. However, the situation can be understood quite easily by reference to Fig. 36. The main mirror is

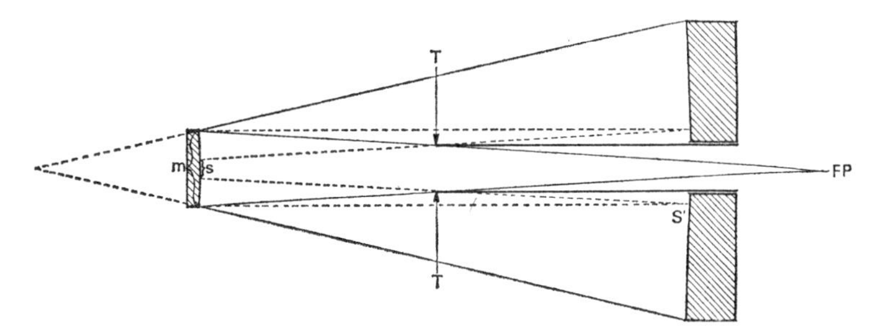

Fig. 36 Designing a shade tube for a Cassegrain telescope

imaged in the secondary as a disc of diameter m, which corresponds to the width of the converging cone from the primary at the position of the secondary. If FP marks the position of the axial focal point, a diaphragm on the cone formed from m to FP will exclude all light except that contributing to the axial image. If we consider this diaphragm as being formed by the front of the shade tube, we now need to know how long and how wide the tube must be.

The secondary casts a shadow on the primary equal to its own diameter s and the image of this imaginary shadow in the secondary has a diameter of s^1, being minified in the same ratio as the image of the primary. Drawing a straight line between the edge of s and the edge of s^1 indicates the outer permissible limit of the shade tube's diameter. If it anywhere exceeds this limit, the effect is to increase the apparent diameter of the central obstruction. The crossing-point of this line with the focal cone indicates the desired length and diameter of the shade tube (T).

If we are interested only in axial images, the diameter of the secondary need be no more than the calculated value of m. This will mean that the secondary will not be large enough to reflect the full image of the primary to all parts of the field of view, but the effect will be unnoticeable at the sort of powers used on the typical Cassegrain telescope, where very low-power eyepieces and wide fields of view are impractical because of the long focal length. However, an over-size secondary is of service in acting as a sky-flood shield to those parts of the field of view away from the axis, and it is usual to make the diameter of the secondary about 10 per cent larger than *m*. The diameter of the hole in the primary will be just large enough to allow the shade tube to pass through without being in contact with the glass.

Off-axis reflectors

We have already seen that the central obstruction due to the diagonal in a Newtonian reflector has the effect of brightening the diffraction rings, so lowering the contrast of planetary detail. The same is true, of course, of the secondary mirror in a Cassegrain telescope. If the mirror is kept small (ideally, less than 1/5th of the aperture of the telescope), these effects are almost negligible, but the fundamental solution is to tilt the primary mirror so that the flat (if a Newtonian) or the secondary (if a compound reflector) is altogether clear of the incoming light. However, the primary must be figured to an off-axis shape in order to give good definition, and such work is not for the average amateur. By using mirrors of large focal ratio, and hence spherical, a moderate tilt can be imparted before astigmatism becomes obtrusive, and coma is altogether absent from spherical systems. An off-axis Cassegrain-type telescope, using spherical mirrors, known as the *Schiefspiegler*, is claimed by European amateurs to give the pure definition of the refractor without the drawback of secondary spectrum, although at the expense of a large focal ratio and a restricted field of view.

Catadioptric telescopes: the Schmidt system

The most exciting developments in telescope design for nearly two centuries started in 1930, when the German optician Bernhard Schmidt (1879–1935) produced a system in which the aberrations of a spherical mirror were corrected by placing in front of it a large, thin lens. This lens, or *Schmidt plate*, has to be figured to an undulating shape (Fig. 37), the central

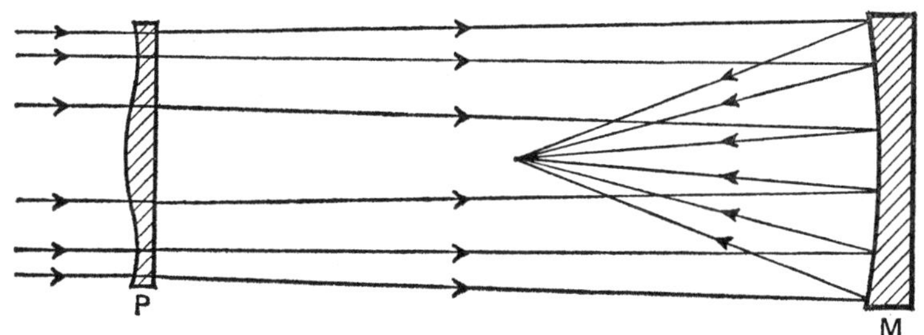

Fig. 37 The Schmidt telescope. M: spherical mirror. P: corrector plate. In a Schmidt camera *the mirror would be considerably larger than the plate, in order to catch off-axis beams*

region converging the rays slightly and the margins diverging them, so balancing the spherical aberration of the mirror. If this plate is placed at the centre of curvature of a spherical mirror, off-axis rays are brought to almost as perfect a focus as axial ones, since they all strike the mirror normally. The only degradation of the marginal images is caused by the off-axis effect of the plate itself, and this is very small. Schmidt designed his system to be used as a wide-angle camera, and fields of excellent definition of 10° or more, at focal ratios down to f/2 or even faster, are possible; the acceptable field of view of an f/2 paraboloid is only a few minutes of arc across. The focal surface of the camera is curved, but the film can quite easily be pulled to the required shape against a convex former.

The Maksutov telescope

The Schmidt idea can be adapted to Newtonian or Cassegrain telescopes to give much wider fields of good definition than are possible with classical aspherised primaries, but the plate is undoubtedly difficult to figure because of its curious shape. Of much greater practical interest to amateur opticians is the modification of Schmidt's scheme by D. D. Maksutov, in which the plate's function is taken over by a steeply-curved, slightly diverging meniscus lens (Fig. 38), which has

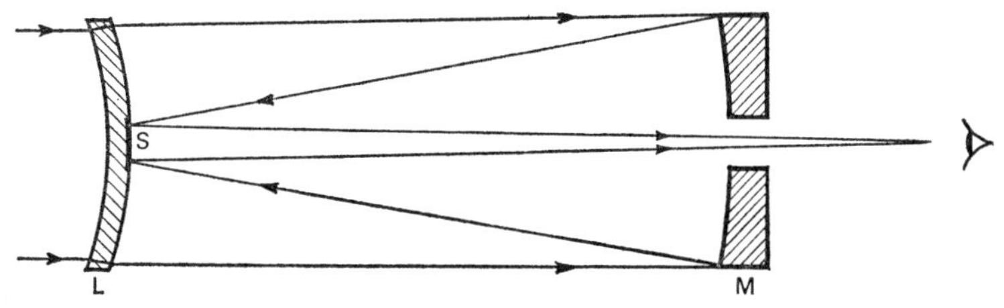

Fig. 38 The Maksutov telescope. M: spherical mirror. L: meniscus corrector shell. S: reflective spot on the shell to act as a Cassegrain secondary mirror

strong negative spherical aberration. This lens or 'shell' has almost spherical surfaces, and so is much easier to figure than a Schmidt plate, even when used with primaries as fast as f/2. At f/4 or slower the system can be used in the Newtonian form, but the vast majority of Maksutovs are in the Cassegrain form, with the secondary's function performed by an aluminised central spot on the second (convex) surface of the shell.

It can be said that the Maksutov telescope has ushered in the age of truly compact, high-performance instruments, and, particularly in the U.S.A., it has been accepted with enthusiasm in amateur projects. Here we have the makings of an ultra-short instrument working at an effective focal ratio of between 12 and 20, the tube sealed against dust and the 'secondary' requiring no separate support, so that diffraction effects around vanes and arms are removed. The shell imparts negligible colour (far less than the equivalent object glass), giving a truly achromatic image. The short tube, so free from vibration effects, makes the telescope a joy to mount and use.

A Maksutov design for the amateur

The Maksutov is not an easy telescope to make because of the precision with which the shell must be ground, but this difficulty is more one of craft than art, and the task is straightforward compared with the problem of figuring the f/2 or f/3 mirror whose aberrations it corrects; at the same time, we save having to make a separate secondary mirror! Another and more serious difficulty encountered by amateurs in the British Isles is the lack of available information. This must be the fundamental reason why so few Maksutovs have been made in this country, and the rest of this chapter is aimed at giving the interested amateur sufficient details to get him started.

The shell

A piece of glass worked optically flat on both sides and then curved to the appropriate radius would impart sufficient negative spherical aberration to a beam to neutralise the positive aberration of a spherical mirror. It would not, however, be achromatic. To remove the unwanted colour, the plate has to be slightly thicker at the centre than at the edge, the relationship being given roughly by

$$t = \frac{d \times N^2}{N^2 - 1}$$

where t is the central thickness of the shell, d is the difference between the radii of curvature of the two sides ($r_2 - r_1$), and N is the refractive index. If the shell were indeed equivalent to a bent parallel-sided plate, then d would of course equal t; in practice, d is always smaller than t. Given this relationship between r_1 and r_2, it is possible to calculate the radius of curvature (r_3) of the spherical mirror whose aberration the plate will combat, and hence to arrive at a design for the whole telescope. Appendix B gives a recommended design from which we now extract the shell details.

Maksutov shell

Diameter 16cm
Central thickness 1.22cm
r_1 −16.355cm
r_2 −17.08cm
N 1.523

The glass chosen for the shell, ophthalmic crown, is plentiful and can be purchased in discs thick enough for the shell to be ground out. However, an enormous amount of labour would be involved, since the depth of curve of the two sides is about 2cm, and a preferable way is to purchase a moulded blank and to carry on the grinding from there. A convenient way of making tools is to mould them, on the rough blank, in plaster of paris. When set, the plaster moulds are lacquered to make them waterproof and glass facets about 4 mm thick and 20mm square are stuck to their faces, using hard pitch. Two more moulds are later coated with pitch to act as polishers.

Measuring the shell

Upon the accuracy of the shell depends the success of the telescope, and the optician has to bring his two radii, and the thickness, all to truth at the same time. The central thickness can be checked directly with a pair of calipers, the gap of which can be measured using a vernier gauge. A tolerance of ±0.1mm should be aimed at, and to make sure that the mark is not overshot the recommended technique is to smooth one side completely before doing more than rough-grinding the other.

There is another advantage in doing this. The limited literature available on the subject makes much of the difficulty of matching the radii of curvature of the two surfaces, for the perfect balancing of the aberrations depends on the difference between them being correct to within about 1 per cent. This tolerance also applies to the thickness, since the two are connected by the formula given above. The difference between the radii in our example being 7.25mm, it appears at first sight as if we must measure the two radii to within 0.04mm or so, and elaborate optical ways of doing this have been described, no doubt putting many people off the project from the outset! However, all we are really interested in is the *difference* between the radii to fractions of a millimetre; as long as they match to within these limits, the radii themselves can wander by about 1 per cent of their whole amount, or $\pm 1\frac{1}{2}$mm, an amount relatively easy to measure, and the telescope will still be perfectly achromatic and well corrected for primary spherical aberration.

To keep track of the difference between the radii of curvature, all that is needed is an accurate spherometer. The writer's (Plate 10) was made from a 0–25mm micrometer reading to 0.001mm on the vernier, and has proved adequate for making Maksutov shells of from 9cm to 27.5cm in aperture – the only other requirement is a tape measure! The anvil casting was sawn off and the base turned circular. A light alloy disc about 90mm in diameter and 30mm thick was then set in the lathe and turned to the shape shown in Fig. 39, so that it could accommodate being set on steeply convex surfaces if desired. Without altering the setting, a central hole was bored to accept the base of the micrometer and a groove indicating a circle about 75mm in diameter was run around the base. Three ball bearings were cemented at 120° intervals around this groove, and the micrometer was secured with a locking screw. Another ball bearing was cemented, via a short sleeve, to the micrometer shaft; it would have been better still to have used a special ball-ended type of micrometer.

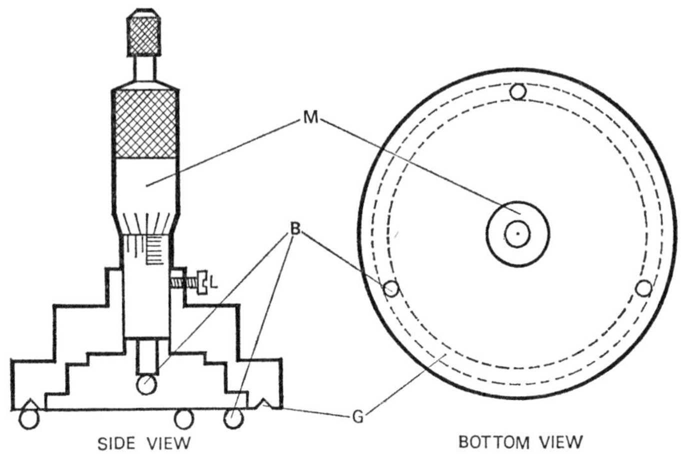

Fig. 39 An accurate spherometer. B: steel balls. G: accurate circular groove to locate balls. M: micrometer. L: locking screw. As shown, the spherometer is set to read on a steeply convex surface

Such an instrument is capable of measuring to within about two wavelengths of yellow light, and so it must be zeroed on a reasonably accurate optical flat. Formulae for accurate sagittal measurements are given in Appendix B. The writer would strongly advise against making the full-diameter spherometers sometimes advocated. They are practically useless for any other work, and are needlessly accurate. The small one described here has the great advantage of being able to measure the radius at different zones of the shell, so detecting any irregular grinding; and it will be of use in practically any other project the worker undertakes.

Let us now suppose that both the surfaces have been rough-ground, with plenty of thickness left over. The concave surface is now taken right down through the grades, noting how much thickness is taken off at each stage to give some help in planning the work on the convex side. Checking can be done with the spherometer, or with a glass template. When finished, a quick polish is given to allow a definitive measurement to be made, using a bare torch bulb. Hold a small screen so that it and the lamp filament are equidistant from the concave surface, and focus the image of the filament sharply.

Measurement of the radius to within 1mm or better will not be difficult if the average of several readings is taken. The spherometer reading on the surface is carefully noted, once again taking the average of several settings, and the new setting for the *measured* value of $r_1+7.25$mm is calculated. The fact that there may be an appreciable error in the measurement of r_1 does not affect the accuracy of the result to any perceptible extent, since we are concerned with differences rather than absolute values. The second surface is then worked until the reading on the concave tool is in agreement.

The task of grinding down the second surface involves keeping an eye on radius, thickness, and wedge. Wedge can be measured with an ordinary micrometer located against the edge of the convex surface, but be most careful to prevent stray particles of grit from being crunched against the glass. The edge thickness should be constant to within about 0.005mm. Because of the possibility of scratches occurring in the fine-smoothing process and having to be ground out, it is advisable to aim near the upper limit on thickness. The following amounts of glass are likely to be removed during the finer stages, assuming that no extra scratch- or pit-removing operations are needed:

220 carborundum	0.12mm
320 carborundum	0.08mm
225 emery	0.03mm
125 emery	0.02mm

This means that a central thickness of about 12.5mm at the conclusion of the 120 work should be safe.

Aim at having the curve of the tool to truth by the time the 220 stage is reached. It will wander during the subsequent work, but only by small amounts, and can be checked and corrected by reversed grinding. If, for any reason, either d or t should finish out of tolerance, the effect can be minimised by altering the other factor by the same amount.

At some stage during the fine-grinding process, a narrow flat margin is ground on the first curve, using grade 220 and

320 carborundum and a sheet of flat glass, to reduce its operating aperture to the required value. This flat edge will be needed to locate the shell in its cell.

Figuring and checking

The rest of the optical work calls for little comment, the polishing of the shell and the grinding of the primary being accomplished by normal techniques. The optical system must be tested as a complete unit in collimated light, preferably supplied by a 20cm or larger paraboloid of known quality; and the primary, having been checked for sphericity, needs to be coated before figuring commences. Residual zones are then removed by working the front (or back) surface of the shell. If all the surfaces have been correctly worked, the knife-edge appearance will show a raised intermediate zone on the shell (in other words, a zone of short focus), and an oblating stroke with a small polisher will be needed. Once a satisfactory figure has been achieved, the shell can be sent away for coating the convex side with an aluminised spot of the required diameter.

An accurately-constructed Maksutov of the size described should show no colour at all in the focused image. The shell is effectively a very weak lens working at about f/100, and it is so well achromatised that the distance between the yellow and red-blue foci is about 0.00004 of the focal length, compared with typical values of about 10 times this for normal achromatic objectives. At larger apertures the very faintest breath of colour – a washed-out, watery tint – may be detected with the knife-edge, but any residual colour visible in a 15cm Maksutov will be the result of poor workmanship.

Conclusion

A final word of advice to the maker of any telescope, whether refractor, Newtonian, or a compound design: shun compromise types. The writer has encountered many people who admire the compactness of the Cassegrain telescope but want to make it adaptable for low-power, wide-field work, either by having an alternative Newtonian diagonal to replace the secondary, or making the amplification low to give only a moderate effective focal length. Both alternatives are bad telescopically for different reasons; it would be far better to make two telescopes, one for wide fields and the other for high powers. Astronomical telescopes are specialised instruments, and work better if they are treated as such. The maker should always have a very clear idea of the work he intends doing before embarking on the project, since this will enable him to rationalise his design alternatives before the point of no return is reached.

APPENDIX A

Suppliers of Materials

Abrasives, pitch, etc	Brunnings (Holborn) Ltd., 133 High Holborn, London WC1 David Hinds Ltd., 2 Wolsey Road, Hemel Hempstead, Herts.
Aluminising	David Hinds Ltd. Optical Works Ltd., 32 The Mall, Ealing, London W5
Books	Astro Books & Supplies, 342 Lower Addiscombe Road, Croydon, Surrey Geoffrey Falworth, 11 Wimbledon Avenue, Blackpool, Lancs.
Eyepieces, small lenses, focusing mounts, etc	Astro Books & Supplies H. W. English, 469 Rayleigh Road, Hutton, Brentwood, Essex (Ex-government equipment)
Glass	Chance-Pilkington Ltd., Glascoed Road, St Asaph, Flintshire (Optical) T. & W. Ide Ltd., Glasshouse Fields, London E1 (Plate) H. V. Skan Ltd., 425–433 Stratford Road, Shirley, Solihull, Warwickshire (Agents for Schott optical and low-expansion glass)
Mirror kits	David Hinds Ltd.

Mirrors and object glasses	Astronomical Equipment Ltd., Lea Industrial Estate, Ox Lane, Harpenden, Herts. David Hinds Ltd. H. Wildey, 14 Savernake Road, Hampstead, London NW3
Telescope manufacturers	Astronomical Equipment Ltd. Bedford Astronomical Supplies, 5B Old Bedford Road, Luton, Beds. (Catadioptric telescopes) Fullerscopes, 63 Farringdon Road, London EC1 H. N. Irving & Son, 258 Kingston Road, Teddington, Middlesex
Variable-frequency oscillators	Astrotech, 39 Periwinkle Lane, Dunstable, Beds.

APPENDIX B

Optical Formulae and Designs

The beginner in astronomical telescope making will require no formulae beyond those already given in the text. But the writer has often found difficulty in locating formulae for more advanced work, and the ones given below will be found useful by anyone attempting telescopes that go a stage beyond the simple Newtonian.

Aberrations. The longitudinal spherical aberration, Δf, of a mirror or lens of focal length f and aperture D, is given by

$$\Delta f = k \times \frac{D^2}{4}$$

For a spherical mirror, $k=+0.125$. The value of a lens depends upon the design.

The longitudinal chromatic aberration of a single lens is given by

$$\Delta f = \frac{f}{V}$$

where V is the dispersive power of the glass used, and indicates the distance between the foci of the colours for which the value of V applies.

Cassegrain telescope. The relationship between the radii of curvature and relative positions of the mirrors in a Cassegrain telescope are as follows.

F=Focal length of primary
r=Radius of curvature of secondary
A=Amplification of secondary

P=Distance of secondary inside prime focus
P_1=Distance from secondary to final focus
f/=Focal ratio of primary
b=Distance of final focus behind primary
d=Minimum necessary diameter of secondary

Then:

$$r = \frac{2\,P\,P_1}{P_1 - P},\quad A = \frac{P_1}{P},\quad P = \frac{F + b}{A + 1},\quad d = \frac{P}{f/}$$

The eccentricity e of the secondary, assuming the primary to be parabolic, is given by

$$e = \frac{A + 1}{A - 1}$$

Dall-Kirkham telescope. The basic formulae are the same as for the Cassegrain. To find the required eccentricity of the primary to balance the aberration of the spherical secondary, use

$$e = \sqrt{1 - \left[\frac{2P^2}{Fr}\left(\frac{P + P_1}{P_1}\right)^2\right]}$$

The conjugates for testing the primary are obtained from the ellipsoid formula. The value of e for a typical Dall-Kirkham system is about 0.9.

Diffraction discs. The theoretical diameter of the diffraction image of a star is given by

$$s = 2.44\lambda \times f/$$

where λ is the wavelength of light used, and f/ is the focal ratio of the telescope's objective. In other words, *any* f/10 telescope of perfect quality will give a diffraction disc 24.4λ across (about 0.012mm for yellow light). The *angular* diameter of the disc, in yellow light, is obtained by

$$s = \frac{28''}{D}$$

where D is the aperture of the objective. Because the brightness of the diffraction disc falls off rapidly towards its margin, the true edge cannot normally be observed, and the practical telescopic diameter of a star image is given more closely by 11.5″/D.

Ellipsoid conjugates. For a prolate ellipsoid of eccentricity e and mean radius of curvature R, the distances of the two null conjugates from the mirror are $\frac{R}{1+e}$ and $\frac{R}{1-e}$.

Foucault test. The knife-edge movement from the central position ($z=0$) to the centre of curvature of any zone of radius z, on a concave surface of eccentricity e and mean radius of curvature R, is given by

$$\Delta R = \frac{z^2 e^2}{R}$$

The movement is towards the mirror if the surface is oblate (e negative), and away from it if the surface is prolate (e positive).

Sagitta. The following formulae are used when trying to evaluate the radius R of a surface in terms of the sagittal reading S of a spherometer of radius r, or vice versa.

$$R = \frac{r^2}{2S} + \frac{S}{2}$$

or

$$S = R - \sqrt{R^2 - r^2}$$

$$= \frac{r^2}{2R} + \frac{r^4}{8R^3} + \frac{r^6}{16R^5} \dots$$

The term alone is accurate to about 0.5 per cent when the measured sagitta is less than 1 per cent of R, or when the diameter of the spherometer is less than about 0.25R.

If the spherometer has ball-type feet of diameter b, the derived value of R must be corrected by *adding* b/2 if the surface is *concave,* and *subtracting* b/2 if it is *convex.* The complete expression for finding the precise radius of curvature of any surface would therefore be

$$R = \frac{r^2}{2S} + \frac{S}{2} \pm \frac{b}{2}$$

Schmidt telescope. The correcting plate for a Schmidt system usually has one side flat and the asphericity applied to the other side. If the greatest depth of curve is near the 71 per cent zone – which is the most convenient form to figure – the value of k in the following formulae is 1.0. This formula gives the variation of thickness of the plate along a radius for any zone z, where r is the radius of the corrector plate, N is the refractive index of the glass, and R is the radius of curvature of the spherical primary mirror at whose centre of curvature the plate is placed:

$$t = \frac{z^4 - kr^2z^2}{4(N-1)R^3}$$

To find the maximum depth of curve, as an indication of the amount of glass to be removed, use

$$t_{max} = \frac{k^2r^2}{16(N-1)R^3}$$

Optical designs. The following design for a 20cm f/4–20 Dall-Kirkham telescope uses the symbols already given.

D=20cm
d=3.96cm (recommended diameter 4.4cm)
R=2F=160cm
r=39.57cm
P=15.83cm
P_1=79.17cm
b=15cm
Diameter of perforation=3cm (approx.)
Primary-secondary separation=64.17cm

The eccentricity e of the primary is 0.88, and the conjugates for null testing are 85.17cm and 1318cm. To avoid the possibility of flexure in working, the disc from which the secondary is made should be about 8mm thick.

The following design for an f/2.5–14 Maksutov-Cassegrain is based on one by the writer that has been tested and proved in commercial practice.

Shell

Glass type: ophthalmic crown (N=1.523, V=59.0)
Overall diameter 16cm
Clear front aperture 15.00cm
Clear back aperture 15.42cm
Central thickness 1.22cm
Radius of curvature, 1st surface −16.355cm
Radius of curvature, 2nd surface −17.08cm
Diameter of aluminised spot 3.3cm

Mirror

Diameter 16cm
Radius of curvature −73.63cm
Diameter of perforation 2.5cm (approx.)
Mirror-shell separation 30.49cm (the final focus falling 11cm behind the mirror)

The theoretical diameter of the primary is 15.75cm, but it is advisable to make it somewhat oversize to ensure that the full aperture of the corrector shell is being used.

In addition to the two object glass designs mentioned in the text, the following one, by James G. Baker, offers superior spherical correction and negligible coma. Despite this, we still have the convenient situation where the second and third surfaces are of equal and opposite curvature, so that they can be worked on one another. However, the fourth surface is convex, though only weakly so.

The design is of the 'air-spaced' type, and a short, accurately-made, cylindrical spacer will be needed to give the correct separation between the centres of the second and third surfaces. This, and the cell itself, will have to be made with a lathe.

The original design was for glasses made in the U.S.A., but the Schott glasses BK7 and F4 differ negligibly from the design specification.

Crown: N=1.517, V=64.2
Flint: N=1.617, V=36.6
$r_1=+0.5814$
$r_2=-0.3585$
$r_3=-0.3585$
$r_4=-1.6189$
Air-space between centres 0.00311

The radii are given in terms of the focal length of the finished lens. The central thicknesses of the lenses are also of importance, and should be within a few per cent of 0.00693 and 0.00495 for the crown and flint elements respectively. (Design from *Telescopes and Accessories;* see Appendix C.)

APPENDIX C

Bibliography and Societies

Books about Telescopes

G. Z. Dimitroff & J. G. Baker: *Telescopes & Accessories* (J. & A. Churchill, London, 1947). This scarce book is well worth seeking out for Baker's designs and the general optical information it contains.

N. E. Howard: *Handbook for Telescope Making* (Faber & Faber, London, 2nd edition 1969). British printing of a popular American introduction.

A. G. Ingalls (ed.): *Amateur Telescope Making* (Scientific American Inc., U.S.A., 1953). The standard work, in three volumes that gradually appeared over a quarter of a century. Contains a tremendous amount of information, but much of it is difficult to locate; mostly American, and somewhat overwhelming for the beginner. Much has happened in amateur telescope-making since 1953, and a fourth volume would be welcome.

H. C. King: *The History of the Telescope* (Charles Griffin, London, 1955). An encyclopaedic study. Although not a practical book, the reader will learn much from discovering the methods and difficulties of the great telescope-makers of history.

J. B. Sidgwick: *Amateur Astronomer's Handbook* (Faber & Faber, London, 3rd edition 1971). This important work studies telescope, observer, and observational methods, and comes near to being indispensable to the practical amateur.

Maksutov Articles (1963). This 39-page compendium represents the most comprehensive publication currently available on amateur Maksutov telescopes. It consists of reprints from the world-circulation astronomical magazine *Sky & Telescope,*

and can be obtained from its offices, 49–51 Bay State Road, Cambridge, Mass. 02138, U.S.A.
J. Texereau: *How to Make a Telescope* (Interscience Publishers, New York, 1963). A translation of the book by the leading French optician, and widely referred to.

Astronomical Societies
The two national amateur societies in the British Isles are the British Astronomical Association (Burlington House, Piccadilly, London W1), and the Junior Astronomical Society (58 Vaughan Gardens, Ilford, Essex). The B.A.A. includes some of the country's leading telescope makers in its ranks.

There are a great number of local societies, and joining one represents the best course of action for the beginner, since he will find it easy to meet fellow-enthusiasts. A list of British societies can be obtained by sending a stamped addressed envelope to the Federation of Astronomical Societies, 130 Derinton Road, Tooting, London SW17.

Index